AS ... at.
To do we ... se hard.

Th ... d,

For ... ons.

And of course, we've done our best to make the whole thing vaguely entertaining for you.

Complete Revision and Practice

Exam Board: Edexcel

Editors:
Mary Falkner, Sarah Hilton, Paul Jordin, Sharon Keeley, Simon Little, Andy Park.

Contributors:
Antonio Angelosanto, Vikki Cunningham, Ian H. Davis, John Duffy, Max Fishel, Emma Grimwood, Richard Harwood, Philippa Hulme, Lucy Muncaster, Glenn Rogers, David Scott, Derek Swain, Paul Warren, Chris Workman.

Proofreaders:
Barrie Crowther, Julie Wakeling.

Published by CGP

ISBN: 978 1 84762 124 5

With thanks to Jan Greenway for the copyright research.

With thanks to Science Photo Library for permission to reproduce the photograph used on page 108.

Graph to show trend in atmospheric CO_2 Concentration and global temperature on page 113 based on data by EPICA Community Members 2004 and Siegenthaler et al 2005.

Groovy website: www.cgpbooks.co.uk
Jolly bits of clipart from CorelDRAW®
Printed by Elanders Ltd, Newcastle upon Tyne.

Based on the classic CGP style created by Richard Parsons.

Contents

How Science Works

Unit 1

Section 1 — Formulas and Equations

Section 2 — Energetics

Section 3 — Atomic Structure

Section 4 — Bonding

Section 5 — Organic Chemistry

Unit 2

Section 1 — Bonding & Intermolecular Forces

Section 2 — Inorganic Chemistry

Section 3 — Kinetics and Equilibria

Section 4 — Organic Chemistry 2

Section 5 — Mechanisms, Spectra & Green Chemistry

The Scientific Process

'How Science Works' is all about the scientific process — how we develop and test scientific ideas. It's what scientists do all day, every day (well except at coffee time — never come between scientists and their coffee).

Scientists Come Up with **Theories** — Then **Test Them**...

Science tries to explain **how** and **why** things happen. It's all about seeking and gaining **knowledge** about the world around us. Scientists do this by **asking** questions and **suggesting** answers and then **testing** them, to see if they're correct — this is the **scientific process**.

1) **Ask** a question — make an **observation** and ask **why or how** whatever you've observed happens.
   E.g. Why does sodium chloride dissolve in water?
2) **Suggest** an answer, or part of an answer, by forming a **theory** or a **model** (a possible **explanation** of the observations or a description of what you think is happening actually happening).
   E.g. Sodium chloride is made up of charged particles which are pulled apart by the polar water molecules.
3) Make a **prediction** or **hypothesis** — a **specific testable statement**, based on the theory, about what will happen in a test situation.
   E.g. A solution of sodium chloride will conduct electricity much better than water does.
4) Carry out **tests** — to provide **evidence** that will support the prediction or refute it.
   E.g. Measure the conductivity of water and of sodium chloride solution.

The evidence supported Quentin's Theory of Flammable Burps.

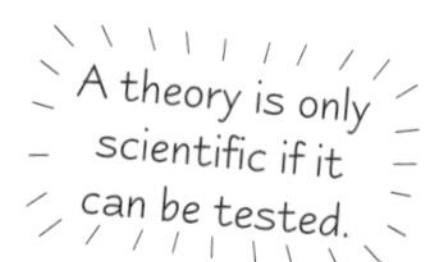

...Then They **Tell** Everyone About Their **Results**...

The results are **published** — scientists need to let others know about their work. Scientists publish their results in **scientific journals**. These are just like normal magazines, only they contain **scientific reports** (called papers) instead of the latest celebrity gossip.

1) Scientific reports are similar to the **lab write-ups** you do in school. And just as a lab write-up is **reviewed** (marked) by your teacher, reports in scientific journals undergo **peer review** before they're published.
   Scientists use standard terminology when writing their reports. This way they know that other scientists will understand them. For instance, there are internationally agreed rules for naming organic compounds, so that scientists across the world will know exactly what substance is being referred to. See page 48.
2) The report is sent out to **peers** — other scientists who are experts in the **same area**. They go through it bit by bit, examining the methods and data, and checking it's all clear and logical. When the report is approved, it's **published**. This makes sure that work published in scientific journals is of a **good standard**.
3) But peer review **can't guarantee** the science is **correct** — other scientists still need to **reproduce** it.
4) Sometimes **mistakes** are made and bad work is published. Peer review **isn't perfect** but it's probably the best way for scientists to self-regulate their work and to publish **quality reports**.

...Then **Other Scientists** Will **Test** the Theory Too

1) Other scientists read the published theories and results, and try to **test the theory** themselves. This involves:
   - Repeating the **exact same experiments**.
   - Using the theory to make **new predictions** and then testing them with **new experiments**.
2) If all the experiments in the world provide evidence to back it up, the theory is thought of as **scientific 'fact'** (for now).
3) If **new evidence** comes to light that **conflicts** with the current evidence the theory is questioned all over again. More rounds of **testing** will be carried out to try to find out where the theory **falls down**.

This is how the **scientific process works** — **evidence** supports a theory, loads of other scientists read it and test it for themselves, eventually all the scientists in the world **agree** with it and then bingo, you get to **learn** it.

This is exactly how scientists arrived at the structure of the atom (see page 4) — and how they came to the conclusion that electrons are arranged in shells and orbitals (see page 28). It took years and years for these models to be developed and accepted — this is often the case with the scientific process.

The Scientific Process

If the **Evidence** Supports a Theory, It's **Accepted — for Now**

Our currently accepted theories have survived this '**trial by evidence**'. They've been tested **over and over again** and each time the results have backed them up. **BUT**, and this is a big but (teehee), they never become totally indisputable fact. Scientific **breakthroughs or advances** could provide new ways to question and test the theory, which could lead to **changes and challenges** to it. Then the testing starts all over again...

And this, my friend, is the **tentative nature of scientific knowledge** — it's always **changing** and **evolving**.

When CFCs were first used in fridges in the 1930s, scientists thought they were problem-free — well, why not? There was no evidence to say otherwise. It was decades before anyone found out that CFCs were actually making a whopping great hole in the ozone layer. See page 108-109.

Evidence Comes From **Lab Experiments...**

1) Results from **controlled experiments** in **laboratories** are **great**.
2) A lab is the easiest place to **control variables** so that they're all **kept constant** (except for the one you're investigating).
3) This means you can draw meaningful **conclusions**.

For example, if you're investigating how temperature affects the rate of a reaction you need to keep everything but the temperature constant, e.g. the pH of the solution, the concentration of the solution, etc. (See pages 92-93.)

...But You **Can't** Always do a Lab Experiment

There are things you **can't** study in a lab. And outside the lab controlling the variables is tricky, if not impossible.

- *Are increasing CO_2 emissions causing climate change?*
  There are other variables which may have an effect, such as changes in solar activity. You can't easily rule out every possibility. Also, climate change is a very **gradual process**. Scientists won't be able to tell if their predictions are correct for donkey's years.
- *Does drinking chlorinated tap water increase the risk of developing certain cancers?*
  There are always differences between groups of people. The best you can do is to have a **well-designed study** using **matched groups — choose two groups** of people (those who drink tap water and those who don't) which are **as similar as possible** (same mix of ages, same mix of diets etc). But you still can't rule out every possibility. Taking new-born identical twins and treating them identically, except for making one drink gallons of tap water and the other only pure water, might be a fairer test, but it would present huge **ethical problems**.

See pages 112-114 for more on climate change.

Samantha thought her study was very well designed — especially the fitted bookshelf.

Science Helps to Inform **Decision-Making**

Lots of scientific work eventually leads to **important discoveries** that **could** benefit humankind — but there are often **risks** attached (and almost always **financial costs**).

Society (that's you, me and everyone else) must weigh up the information in order to **make decisions** — about the way we live, what we eat, what we drive, and so on. Information is also be used by **politicians** to devise policies and laws.

- **Chlorine** is added to water in **small quantities** to disinfect it. Some studies link drinking chlorinated water with certain types of cancer. But the risks from drinking water contaminated by nasty bacteria are far, far greater. There are other ways to get rid of bacteria in water, but they're heaps **more expensive**.
- Scientific advances mean that **non-polluting hydrogen-fuelled cars** can be made. They're better for the environment, but are really expensive. Also, it'd cost a fortune to adapt the existing filling stations to store hydrogen.
- Pharmaceutical drugs are really expensive to develop, and drug companies want to make money. So they put most of their efforts into developing drugs that they can sell for a good price. Society has to consider the **cost** of buying new drugs — the **NHS** can't afford the most expensive drugs without **sacrificing** something else.

So there you have it — how science works...

Hopefully these pages have given you a nice intro to how science works, e.g. what scientists do to provide you with 'facts'. You need to understand this, as you're expected to know how science works yourselves — for the exam and for life.

The Atom

This stuff about atoms and elements should be ingrained on your brain from GCSE. You do need to know it perfectly though if you are to negotiate your way through the field of man-eating tigers which is AS Chemistry.

Atoms are made up of **Protons, Neutrons** and **Electrons**

All elements are made of **atoms**. Atoms are made up of 3 types of particle — **protons**, **neutrons** and **electrons**.

Electrons
1) Electrons have **–1** charge.
2) They whizz around the nucleus in **orbitals**. The orbitals take up most of the **volume** of the atom.

Nucleus
1) Most of the **mass** of the atom is concentrated in the nucleus.
2) The **diameter** of the nucleus is rather titchy compared to the whole atom.
3) The nucleus is where you find the **protons** and **neutrons**.

The mass and charge of these subatomic particles is **really small**, so **relative mass** and **relative charge** are used instead.

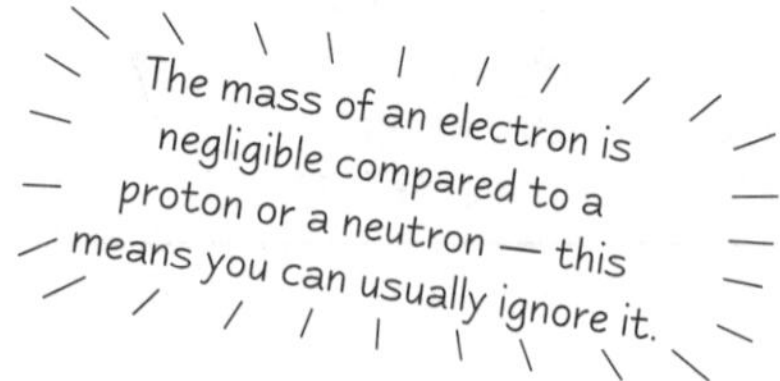

Subatomic particle	Relative mass	Relative charge
Proton	1	+1
Neutron	1	0
Electron, e^-	$\frac{1}{2000}$	–1

Nuclear Symbols Show Numbers of **Subatomic Particles**

You can figure out the **number** of protons, neutrons and electrons from the **nuclear symbol**.

$${}^{A}_{Z}X$$

Mass number
This tells you the **total** number of **protons** and **neutrons** in the nucleus.

Element symbol

Atomic (proton) number
1) This is the number of **protons** in the nucleus — it identifies the element.
2) **All** atoms of the same element have the **same** number of protons.

Sometimes the atomic number is left out of the nuclear symbol, e.g. ^{7}Li. You don't really need it because the element symbol tells you its value.

1) For **neutral** atoms with no overall charge, the number of electrons is **the same as** the number of protons.
2) The number of neutrons is just **mass number minus atomic number**, i.e. 'top minus bottom' in the nuclear symbol.

Nuclear symbol	Atomic number, Z	Mass number, A	Protons	Electrons	Neutrons
${}^{7}_{3}Li$	3	7	3	3	7 – 3 = **4**
${}^{80}_{35}Br$	35	80	35	35	80 – 35 = **45**
${}^{24}_{12}Mg$	12	24	12	12	24 – 12 = **12**

"Hello, I'm Newt Ron..."

Ions have **Different** Numbers of **Protons and Electrons**

Negative ions have **more electrons** than protons...

E.g.

Br^- The negative charge means that there's 1 more electron than there are protons. Br has 35 protons (see table above), so Br^- must have 36 electrons. The overall charge = + 35 – 36 = –1.

...and **positive** ions have **fewer electrons** than protons. It kind of makes sense if you think about it.

E.g.

Mg^{2+} The 2+ charge means that there's 2 fewer electrons than there are protons. Mg has 12 protons (see table above), so Mg^{2+} must have 10 electrons. The overall charge = +12 – 10 = +2.

The Atom

Isotopes are Atoms of the Same Element with Different Numbers of Neutrons

Make sure you **learn** this definition and totally **understand** what it means —

Isotopes of an element are atoms with the same number of protons but different numbers of neutrons.

Chlorine-35 and chlorine-37 are examples of isotopes.

35 – 17 = 18 neutrons ⟵ **Different** mass numbers mean different numbers of neutrons. ⟶ **37 – 17 = 20 neutrons**

$^{35}_{17}Cl$

The **atomic numbers** are the same. **Both** isotopes have 17 protons and 17 electrons.

$^{37}_{17}Cl$

1) It's the **number** and **arrangement** of electrons that decides the **chemical properties** of an element. Isotopes have the **same configuration of electrons**, so they've got **virtually identical** chemical properties.
2) Isotopes of an element do have slightly different **physical properties** though, such as different densities, rates of diffusion, etc. This is because **physical properties** tend to depend more on the **mass** of the atom.

Here's another example — naturally occurring **magnesium** consists of 3 main isotopes.

^{24}Mg (79%)	^{25}Mg (10%)	^{26}Mg (11%)
12 protons	12 protons	12 protons
<u>12</u> neutrons	<u>13</u> neutrons	<u>14</u> neutrons
12 electrons	12 electrons	12 electrons

The periodic table gives the atomic number for each element. The other number isn't the mass number — it's the relative atomic mass (see page 6). They're a bit different, but for all but the most accurate work, you can assume they're equal.

Practice Questions

Q1 Draw a diagram showing the structure of the atom, labelling each part.

Q2 Define the term 'isotope' and give an example.

Q3 Draw a table showing the relative charge and relative mass of the three subatomic particles found in atoms.

Q4 Using an example, explain the terms 'atomic number' and 'mass number'.

Q5 Where is the mass concentrated in an atom, and what makes up most of the volume of an atom?

Exam Questions

Q1 Hydrogen, deuterium and tritium are all isotopes of each other.
  a) Identify one similarity and one difference between these isotopes. [2 marks]
  b) Deuterium can be written as ^{2}H. Determine the number of protons, neutrons and electrons in neutral deuterium. [3 marks]
  c) Write a nuclear symbol for tritium, given that it has 2 neutrons. [1 mark]

Q2 This question relates to the atoms or ions A to D: A. $^{32}_{16}S^{2-}$, B. $^{40}_{18}Ar$, C. $^{30}_{16}S$, D. $^{42}_{20}Ca$.
  a) Identify the similarity for each of the following pairs, justifying your answer in each case.
    (i) A and B. [2 marks]
    (ii) A and C. [2 marks]
    (iii) B and D. [2 marks]
  b) Which two of the atoms or ions are isotopes of each other? Explain your reasoning. [2 marks]

Got it learned yet? — isotope so...

This is a nice straightforward page just to ease you in to things. Remember that positive ions have fewer electrons than protons and negative ions have more electrons than protons. Get this straight in your mind otherwise you'll end up in a right mess. There's nowt too hard about isotopes neither. They're just the same element with different numbers of neutrons.

Atoms and Moles

You have to do all kinds of calculations with masses of atoms, molecules, moles and whatnot. So you need to know which mass to use when, and how masses, moles and concentration are all linked. That's what these pages are for.

Relative Masses are Masses of Atoms Compared to Carbon-12

The actual mass of an atom is **very, very tiny**. Don't worry about exactly how tiny for now, but it's far **too small** to weigh. So, the mass of one atom is compared to the mass of a different atom. This is its **relative mass**. Here are some definitions for you to learn.

The **relative atomic mass**, A_r, is the **average mass** of an atom of an element on a scale where an atom of **carbon-12** is 12.

Relative isotopic mass is the mass of an atom of an **isotope** of an element on a scale where an atom of **carbon-12** is 12.

Relative atomic mass is an average, so it's not usually a whole number. Relative isotopic mass is always a whole number (at AS level anyway). E.g. a natural sample of chlorine contains a mixture of ^{35}Cl (75%) and ^{37}Cl (25%), so the relative isotopic masses are 35 and 37. But its relative atomic mass is 35.5.

The **relative molecular mass**, M_r, is the average mass of a **molecule** on a scale where an atom of **carbon-12** is 12.

For substances that are ionic (or giant covalent, such as SiO_2), relative formula mass is often used instead. You just add up the masses of everything in the formula unit (e.g. NaCl) in exactly the same way.

To find the relative molecular mass, just add up the relative atomic mass values of all the atoms in the molecule, e.g. $M_r(C_2H_6O) = (2 \times 12) + (6 \times 1) + 16 = 46$.

A Mole is Just a (Very Large) Number of Particles

1) Amount of substance is measured using a unit called the **mole** (**mol** for short) and given the symbol ***n***.
2) One mole is roughly **6×10^{23} particles** (**the Avogadro constant, L**).
3) It **doesn't matter** what the particles are. They can be atoms, molecules, penguins — **anything**.
4) Here's a nice simple formula for finding the number of moles from the number of atoms or molecules:

$$\text{Number of moles} = \frac{\text{Number of particles you have}}{\text{Number of particles in a mole}}$$

Example:
I have 1.5×10^{24} carbon atoms.
How many moles of carbon is this?

$$\text{Number of moles} = \frac{1.5 \times 10^{24}}{6 \times 10^{23}} = \mathbf{2.5\ moles}$$

Molar Mass is the Mass of One Mole

Molar mass, M, is the mass of **one mole** of something.
But the main thing to remember is:

Molar mass is just the same as the relative molecule mass, M_r

That's why the mole is such a ridiculous number of particles (6×10^{23}) — it's the number of particles for which the weight in g is the same as the relative molecular mass.

The only difference is you stick a 'g mol^{-1}' for grams per mole on the end...

Example: Find the molar mass of $CaCO_3$.

Relative molecular mass, M_r, of $CaCO_3$ = 40 + 12 + (3 × 16) = 100
So the molar mass, M, is **100 g mol^{-1}**. — i.e. 1 mole of $CaCO_3$ weighs 100 g.

Here's another formula. This one's really important — you need it **all the time**:

$$\text{Number of moles} = \frac{\text{mass of substance}}{\text{molar mass}}$$

Example: How many moles of aluminium oxide are present in 5.1 g of Al_2O_3?

Molar mass of Al_2O_3 = (2 × 27) + (3 × 16) = 102 g mol^{-1}

Number of moles of Al_2O_3 = $\frac{5.1}{102}$ = **0.05 moles**

Atoms and Moles

In a Solution the **Concentration** is Measured in **mol dm^{-3}**

1 dm^3 = 1000 cm^3 = 1 litre

1) The **concentration** of a solution is how many **moles** are dissolved per **1 dm³** of solution. The units are **mol dm^{-3}**.
2) Here's the formula to find the **number of moles**.

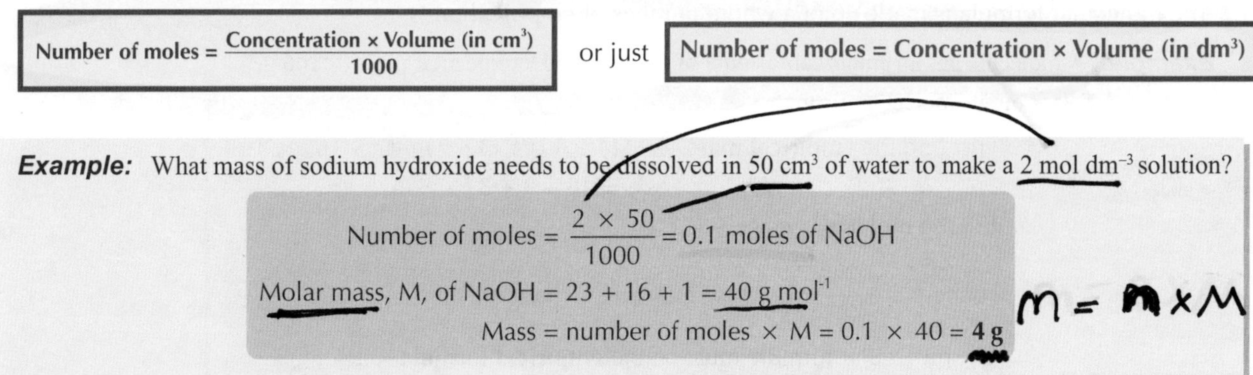

$$\text{Number of moles} = \frac{\text{Concentration} \times \text{Volume (in cm}^3\text{)}}{1000}$$

or just

$$\text{Number of moles} = \text{Concentration} \times \text{Volume (in dm}^3\text{)}$$

Example: What mass of sodium hydroxide needs to be dissolved in 50 cm³ of water to make a 2 mol dm^{-3} solution?

$$\text{Number of moles} = \frac{2 \times 50}{1000} = 0.1 \text{ moles of NaOH}$$

Molar mass, M, of NaOH = 23 + 16 + 1 = 40 g mol^{-1}

Mass = number of moles × M = 0.1 × 40 = **4 g**

3) A solution that has **more moles per dm³** than another is **more concentrated**. A solution that has **fewer moles per dm³** than another is **more dilute**.
4) For really low concentrations, you end up with **tiny** mol dm^{-3} values. They're a bit fiddly, so you're better off using a different unit, such as **parts per million (ppm)** — see below.

Parts Per Million is used for ***Really Small Quantities***

1) The **major gases** in the atmosphere are normally given as **percentages** of the **volume**. But some gases are present in such **tiny amounts** that it's **not very convenient** to write their quantities like this. For instance, **xenon** makes up only **0.000 009%** of the atmosphere. Numbers this small are a pain to work with.
2) So to get round this problem, another type of measurement is used. It is called **parts per million** or **ppm**.
3) So if there's **0.000 009 parts** of xenon in every **one hundred parts of air**, you can multiply both quantities by **10 000** to make the quantity **large enough** to work with, like this:

0.000 009% = 0.000 009 parts per 100 parts of air ⟶ **0.000 009 × 10 000 = 0.09; 100 × 10 000 = 1 000 000** ⟶ **0.09 parts per million**

4) So there's 0.09 ppm xenon. The atmosphere also contains **0.1 ppm** carbon monoxide and **0.3 ppm** nitrous oxide.

Practice Questions

Q1 Explain what relative atomic mass (A_r) and relative isotopic mass mean.

Q2 How many molecules are there in one mole of ethane molecules?

Q3 Write down the equation you'd use to find the number of moles in a certain number of grams of an element.

Q4 Explain how you'd convert a percentage value into parts per million.

Exam Questions

Q1 Calculate the mass of 0.36 moles of ethanoic acid, CH_3COOH. [2 marks]

Q2 What mass of H_2SO_4 is needed to produce 60 cm³ of a 0.25 mol dm^{-3} solution? [2 marks]

You can't pick your relatives, you just have to learn them...

Working out M_r is dead easy — and using a calculator makes it even easier. It'll really help if you know the mass numbers for the first 20 elements or so, or you'll spend half your time looking back at the periodic table. I hope you've done the Practice Questions, cos they pretty much cover the rest of the stuff, and if you can get them right, you've nailed it.

Empirical and Molecular Formulas

Here's another page piled high with numbers — it's all just glorified maths really.

Empirical and *Molecular* Formulas are *Ratios*

1) The **empirical formula** gives just the smallest whole number ratio of atoms in a compound.
2) The **molecular formula** gives the **actual** numbers of atoms in a molecule.
3) The molecular formula is made up of a whole **number** of empirical units.

Example: A molecule has an empirical formula of $C_4H_3O_2$, and a molecular mass of 166 g. Work out its molecular formula.

Compare the empirical and molecular mass.

First find the **empirical mass** — $(4 \times 12) + (3 \times 1) + (2 \times 16) = 48 + 3 + 32 = 83$ g

But the **molecular mass** is 166 g,

so there are $\frac{166}{83} = 2$ empirical units in the molecule.

The molecular formula must be the **empirical formula × 2**,

so the molecular formula = $C_8H_6O_4$. So there you go.

Empirical Formulas are Calculated from *Experiments*

You need to be able to work out empirical formulas from **experimental results** — given as **masses** or **percentages**.

Example 1: When a hydrocarbon is burnt in excess oxygen, 4.4 g of carbon dioxide and 1.8 g of water are made. What is the empirical formula of the hydrocarbon?

First work out how many moles of the products you have.

Note — the only place the carbon in the CO_2 and the hydrogen in the H_2O can have come from is the hydrocarbon.

No. of moles of $CO_2 = \frac{\text{mass}}{M} = \frac{4.4}{12+(16\times2)} = \frac{4.4}{44} = 0.1$ moles

1 mole of CO_2 contains 1 mole of carbon atoms, so you must have started with **0.1 moles of carbon atoms.**

No. of moles of $H_2O = \frac{1.8}{(2\times1)+16} = \frac{1.8}{18} = 0.1$ moles

1 mole of H_2O contains 2 moles of hydrogen atoms (H), so you must have started with **0.2 moles of hydrogen atoms.**

Ratio C : H = 0.1 : 0.2 . Now divide both numbers by the **smallest** — 0.1.
So, the ratio C : H = 1 : 2. So the empirical formula must be CH_2.

Example 2: A compound is found to have percentage composition 56.5% potassium, 8.7% carbon and 34.8% oxygen by mass. Calculate its empirical formula.

If you assume you've got 100 g of the compound, you can turn the % straight into mass, and then work out the number of moles as normal.

Use $n = \frac{\text{mass}}{M}$

In **100 g** of compound there are:

$\frac{56.5}{39} = 1.449$ moles of K $\frac{8.7}{12} = 0.725$ moles of C $\frac{34.8}{16} = 2.175$ moles of O

Divide each number of moles by the **smallest number** — in this case it's 0.725.

K: $\frac{1.449}{0.725} = 2.0$ C: $\frac{0.725}{0.725} = 1.0$ O: $\frac{2.175}{0.725} = 3.0$

The ratio of K : C : O = 2 : 1 : 3. So you know the empirical formula's got to be K_2CO_3.

1) A warning — if the **measurements** you use in these calculations are shoddy, the final answer (the formula) could be wrong. There's always a limit to how **precise** measurements can be and how certain you can be about the result (see page 13). Getting the wrong answer in an AS Chemistry practical isn't the end of the world. But...
2) Say you work for a pharmaceutical company, and you've found a 'miracle substance' in a plant. You'll want to know its chemical formula so you can synthesise the stuff and sell it. You'd certainly want to get the **right formula** in this case, or the pills won't work and you'll have wasted a lot of time and money. And probably get the sack.

Empirical and Molecular Formulas

Molecular Formulas are Calculated from Experimental Data Too

Once you know the empirical formula, you just need a bit more info and you can work out the **molecular formula** too.

Example:

4.6 g of an alcohol, with molar mass 46 g, is burnt in excess oxygen. It produces 8.8 g of carbon dioxide and 5.4 g of water. Calculate the empirical formula for the alcohol and then its molecular formula.

The carbon in the CO_2 and the hydrogen in the H_2O must have come from the alcohol — work out the number of moles of each of these.

No. of moles of $CO_2 = \frac{\text{mass}}{M} = \frac{8.8}{44} = 0.2$ moles

1 mole of CO_2 contains 1 mole of C. So, 0.2 moles of CO_2 contains **0.2 moles of C.**

No. of moles $H_2O = \frac{\text{mass}}{M} = \frac{5.4}{18} = 0.3$ moles

1 mole of H_2O contains 2 moles of H. So, 0.3 moles of H_2O contain **0.6 moles of H.**

Mass of C = no. of moles × M = 0.2 × 12 = 2.4 g
Mass of H = no. of moles × M = 0.6 × 1 = 0.6 g
Mass of O = 4.6 – (2.4 + 0.6) = 1.6 g

Number of moles O = $\frac{\text{mass}}{M} = \frac{1.6}{16} = 0.1$ moles

Now work out the mass of carbon and hydrogen in the alcohol. The rest of the mass of the alcohol must be oxygen — so work out that too. Once you know the mass of O, you can work out how many moles there is of it.

Molar Ratio = C : H : O = 0.2 : 0.6 : 0.1 = 2 : 6 : 1

Empirical formula = C_2H_6O

Mass of empirical formula = (12 × 2) + (1 × 6) + 16 = 46 g

In this example, the mass of the empirical formula equals the molecular mass, so the empirical and molecular formulas are the same.

Molecular formula = C_2H_6O

When you know the number of moles of each element, you've got the molar ratio. Divide each number by the smallest.

Compare the empirical and molecular masses.

Practice Questions

Q1 What's the difference between a molecular formula and an empirical formula?

Exam Questions

Q1 In an experiment to determine the formula of an oxide of copper, 2.8 g of the oxide was heated in a stream of hydrogen gas until there was no further mass change. 2.5 g of copper remained.

Calculate the empirical formula of the oxide. [A_r(Cu) = 63.5, A_r(O) = 16] [4 marks]

Q2 Hydrocarbon X has a molecular mass of 78 g. It is found to have 92.3% carbon and 7.7% hydrogen by mass. Calculate the empirical and molecular formulae of X. [3 marks]

Q3 When 1.2 g of magnesium ribbon is heated in air, it burns to form a white powder, which has a mass of 2 g. What is the empirical formula of the powder? [2 marks]

Q4 When 19.8 g of an organic acid, A, is burnt in excess oxygen, 33 g of carbon dioxide and 10.8 g of water are produced.
Calculate the empirical formula for A and hence its molecular formula, if M_r(A) = 132. [4 marks]

The Empirical Strikes Back...

With this stuff, it's not enough to learn a few facts parrot-fashion, to regurgitate in the exam — you've gotta know how to use them. The only way to do that is to practise. Go through all the examples on these two pages again, this time working the answers out for yourself. Then test yourself on the practice exam questions. It'll help you sleep at night — honest.

Equations and Calculations

Balancing equations'll cause you a few palpitations — as soon as you make one bit right, the rest goes pear-shaped.

Balanced Equations have **Equal Numbers** of each Atom on **Both Sides**

1) Balanced equations have the **same number** of each atom on **both** sides. They're.. well... you know... balanced.
2) You can only add more atoms by adding **whole compounds**. You do this by putting a number **in front** of a compound or changing one that's already there. You **can't** mess with formulas — ever.

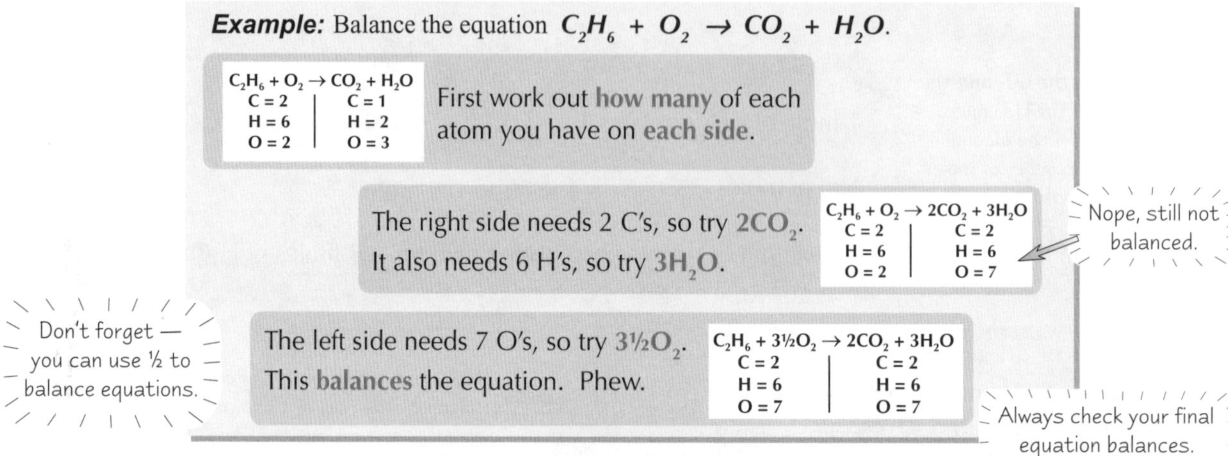

Example: Balance the equation $C_2H_6 + O_2 \rightarrow CO_2 + H_2O$.

$C_2H_6 + O_2$	$\rightarrow CO_2 + H_2O$
C = 2	C = 1
H = 6	H = 2
O = 2	O = 3

First work out how many of each atom you have on each side.

The right side needs 2 C's, so try $2CO_2$. It also needs 6 H's, so try $3H_2O$.

$C_2H_6 + O_2$	$\rightarrow 2CO_2 + 3H_2O$
C = 2	C = 2
H = 6	H = 6
O = 2	O = 7

Nope, still not balanced.

The left side needs 7 O's, so try $3\frac{1}{2}O_2$. This balances the equation. Phew.

$C_2H_6 + 3\frac{1}{2}O_2$	$\rightarrow 2CO_2 + 3H_2O$
C = 2	C = 2
H = 6	H = 6
O = 7	O = 7

Don't forget — you can use ½ to balance equations.

Always check your final equation balances.

In **Ionic Equations** the **Charges** must Balance too

In ionic equations, only the **reacting particles** are included. You don't have to worry about the rest of the stuff.

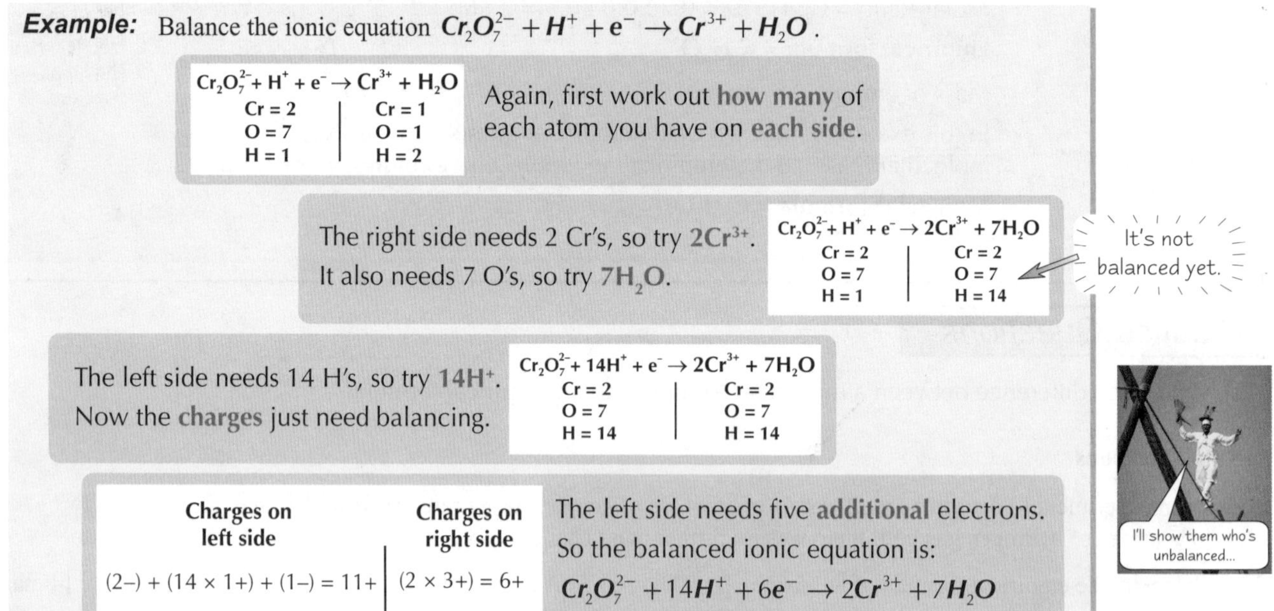

Example: Balance the ionic equation $Cr_2O_7^{2-} + H^+ + e^- \rightarrow Cr^{3+} + H_2O$.

$Cr_2O_7^{2-} + H^+ + e^-$	$\rightarrow Cr^{3+} + H_2O$
Cr = 2	Cr = 1
O = 7	O = 1
H = 1	H = 2

Again, first work out how many of each atom you have on each side.

The right side needs 2 Cr's, so try $2Cr^{3+}$. It also needs 7 O's, so try $7H_2O$.

$Cr_2O_7^{2-} + H^+ + e^-$	$\rightarrow 2Cr^{3+} + 7H_2O$
Cr = 2	Cr = 2
O = 7	O = 7
H = 1	H = 14

It's not balanced yet.

The left side needs 14 H's, so try $14H^+$. Now the charges just need balancing.

$Cr_2O_7^{2-} + 14H^+ + e^-$	$\rightarrow 2Cr^{3+} + 7H_2O$
Cr = 2	Cr = 2
O = 7	O = 7
H = 14	H = 14

Charges on left side	Charges on right side
(2–) + (14 × 1+) + (1–) = 11+	(2 × 3+) = 6+

The left side needs five additional electrons. So the balanced ionic equation is:

$$Cr_2O_7^{2-} + 14H^+ + 6e^- \rightarrow 2Cr^{3+} + 7H_2O$$

I'll show them who's unbalanced...

Balanced Equations can be used to Work out **Masses**

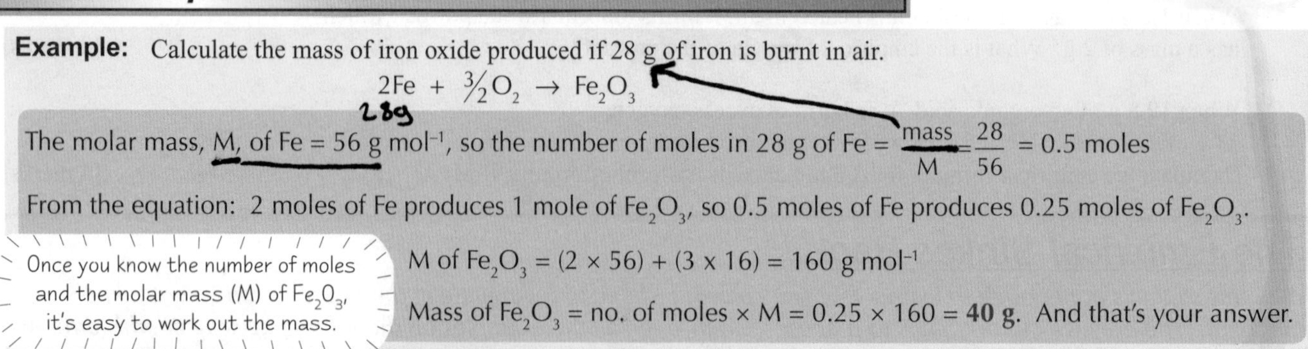

Example: Calculate the mass of iron oxide produced if 28 g of iron is burnt in air.

$$2Fe + \frac{3}{2}O_2 \rightarrow Fe_2O_3$$

28g

The molar mass, M, of Fe = 56 g mol⁻¹, so the number of moles in 28 g of Fe = $\frac{mass}{M} = \frac{28}{56}$ = 0.5 moles

From the equation: 2 moles of Fe produces 1 mole of Fe_2O_3, so 0.5 moles of Fe produces 0.25 moles of Fe_2O_3.

M of Fe_2O_3 = (2 × 56) + (3 × 16) = 160 g mol^{-1}

Mass of Fe_2O_3 = no. of moles × M = 0.25 × 160 = **40 g**. And that's your answer.

Once you know the number of moles and the molar mass (M) of Fe_2O_3, it's easy to work out the mass.

Equations and Calculations

*That's not all... **Balanced Equations** can be used to **Work out Gas Volumes***

If temperature and pressure stay the same, **one mole** of **any** gas always has the **same volume**.
At **room temperature and pressure** (r.t.p.), this happens to be **24 dm³**, (r.t.p is 298 K (25 °C) and 101.3 kPa).
Here are two formulas for working out the number of moles in a volume of gas. Don't forget — **ONLY** use them for r.t.p.

$$\text{Number of moles} = \frac{\text{Volume in dm}^3}{24} \quad \text{OR} \quad \text{Number of moles} = \frac{\text{Volume in cm}^3}{24\,000}$$

Example: How many moles are there in 6 dm³ of oxygen gas at r.t.p.?

$$\text{Number of moles} = \frac{6}{24} = \textbf{0.25 moles of oxygen molecules}$$

It's pretty handy to be able to work out **how much gas** a reaction will produce, so that you can use **large enough apparatus**. Or else there might be a rather large bang.

Example: How much gas is produced when 15 g of sodium is reacted with excess water at r.t.p.?

$$2Na_{(s)} + 2H_2O_{(l)} \rightarrow 2NaOH_{(aq)} + H_{2(g)}$$

Excess water means you know all the sodium will react.

M of Na = 23 g mol⁻¹, so number of moles in 15 g of Na = $\frac{15}{23}$ = 0.65 moles

From the equation, 2 moles Na produces 1 mole H_2,

so you know 0.65 moles Na produces $\frac{0.65}{2}$ = 0.326 moles H_2.

So the volume of H_2 = 0.325 × 24 = **7.8 dm³**

The reaction happens at room temperature and pressure, so you know 1 mole takes up 24 dm³.

*State Symbols** Give a bit More Information about the Substances*

State symbols are put after each compound in an equation. They tell you what **state of matter** things are in.

s = solid	g = gas
l = liquid	aq = aqueous (solution in water)

To show you what I mean, here's an example —

$$CaCO_{3\,(s)} + 2HCl_{(aq)} \rightarrow CaCl_{2\,(aq)} + H_2O_{(l)} + CO_{2\,(g)}$$

solid — solution — solution — liquid — gas

Practice Questions

Q1 What is the state symbol for a solution of hydrochloric acid?

Q2 What is the difference between a full balanced equation and an ionic equation?

Exam Questions

Q1 Calculate the mass of ethene required to produce 258 g of chloroethane, C_2H_5Cl.

$C_2H_4 + HCl \rightarrow C_2H_5Cl$ [4 marks]

Q2 15 g of calcium carbonate is heated strongly until it fully decomposes. $CaCO_{3(s)} \rightarrow CaO_{(s)} + CO_{2(g)}$

a) Calculate the mass of calcium oxide produced. [3 marks]

b) Calculate the volume of gas produced. [3 marks]

Q3 Balance this equation: $KI + Pb(NO_3)_2 \rightarrow PbI_2 + 2KNO_3$ [1 mark]

Don't get in a state about equations...

You're probably completely fed up with all these equations, calculations, moles and whatnot... well hang in there — there's just one more double page coming up. I've said it once, and I'll say it again — practise, practise, practise... it's the only road to salvation (by the way, where is salvation anyway?). Keep going... you're nearly there.

Confirming Equations

To confirm that an equation you've written is correct, you could actually do the reaction — then find out how much of each reactant and product you have and check that it all adds up... always being aware of likely errors.

A Balanced Equation Can Be **Confirmed** by **Experimental Data**

1) A balanced equation tells you **how many moles** of **products** you should expect from given amounts of **reactants**.
2) You can check whether the equation is correct with an **experiment** to **measure** the amount of each product you get. For example, here's a balanced equation for **lithium** reacting with **water**:

$$2Li_{(s)} + 2H_2O_{(l)} \rightarrow 2LiOH_{(aq)} + H_{2(g)}$$

So you'd expect 2 moles of lithium to produce 2 moles of lithium hydroxide and 1 mole of hydrogen.

3) To confirm this equation, you react a **known mass of lithium** with water, and measure **how much lithium hydroxide** and **hydrogen** is actually produced.

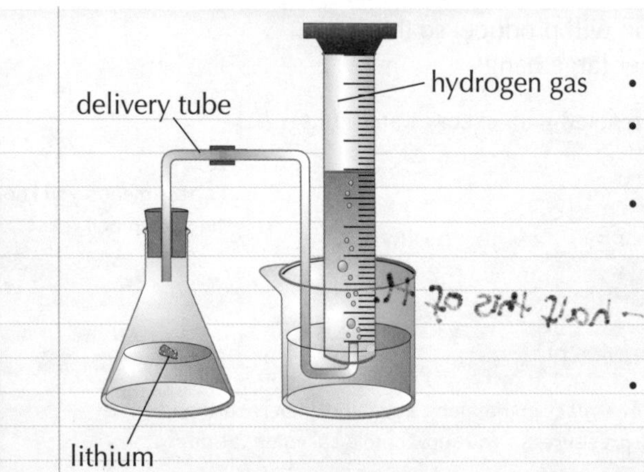

- Weigh out some lithium and drop it into the conical flask.
- Fit the bung and delivery tube into the top of the flask immediately to collect the hydrogen gas as it's produced.
- The **hydrogen** displaces water from the measuring cylinder, so you can read off its volume pretty easily when the reaction has stopped (when the water level stops moving). You're going to use the volume to find the number of moles, so you have to do this at r.t.p. (see page 11).
- The **lithium hydroxide** is produced as a solution in the conical flask. It's an alkali, so to find out how much you've got, you can **titrate it** with an acid (titrations are covered in Unit 2).

The Experimental Data Is Used to **Calculate** the **Number of Moles**

1) To confirm the equation, you need to know how many **moles** of reactants and products there were. It's just a matter of using the right formulas. (See pages 6 and 10-11 for more on these kinds of calculation.)

To find the number of moles of **lithium**, it's this:

$$\text{Number of moles} = \frac{\text{mass of substance}}{\text{molar mass}}$$

And to find the number of moles of **hydrogen gas** (at r.t.p.) you need this one.

$$\text{Number of moles} = \frac{\text{Volume in dm}^3}{24} \quad \text{OR} \quad \text{Number of moles} = \frac{\text{Volume in cm}^3}{24\,000}$$

...And you can find the number of moles of **lithium hydroxide** from the result of a **titration** (page 85 if you want a sneak preview).

2) For example, if **0.14 g of lithium** reacted to produce **222 cm^3** of hydrogen, the number of moles of each are calculated like this:

Number of moles of lithium = 0.14 g ÷ 7 = **0.020 moles**
Number of moles of hydrogen = 222 cm^3 ÷ 24 000 = **0.00925 moles**

This gives you a **moles of lithium : moles of hydrogen** ratio of 2 : 0.925.

3) From the balanced equation for the reaction, you'd expect that ratio to be **2 : 1**. But it's hugely unlikely you'll actually get this **exact answer** from an experiment, however carefully you do it. (See the next page for some possible sources of error.) So an experimental result of 2 : 0.925 is **close enough** to say that the ratio's correct.
4) Then you can work out the ratio 'moles of Li' : 'moles of LiOH'. From the equation, it should be 1 : 1. And if that turns out to be correct too, you can be pretty sure the reaction happened as expected from the balanced equation.

Confirming Equations

The Accuracy of Experimental Data is Always Limited by the Methods Used...

There are several problems with this method that can cause the result to be inaccurate.

1) Lithium has to be stored under **oil**, and it's difficult to **completely remove** the oil — so when you weigh out the lithium, the mass measured may include some oil.
2) When lithium is exposed to the **air** (after you've removed as much oil as possible), it **reacts with oxygen**, forming a surface layer of lithium oxide. So some of the 'mass of lithium' will actually be lithium oxide.
3) Some **hydrogen** will **escape** in the time between dropping the lithium into the flask and fitting the bung and delivery tube — so the volume of hydrogen recorded will be slightly lower than it should be.

These are **systematic errors**, and they're due to the method that's used. They can be **minimised** (by cleaning as much oil from the lithium as possible, etc.) but the only way to **avoid** them is to use a **different method** (which may be impossible without complicated equipment, and might well have problems of its own).

...And the Equipment Used

1) With **all measurements** there's a limit to the **precision** that's possible. For example, a measuring cylinder has a limited number of graduations (markings) on it — the level you're trying to read will usually be somewhere in between two graduations and you'll have to **estimate** its position by eye. The limited precision of equipment leads to **random errors** — sometimes you'll record the value as **greater** than it really is, sometimes as **smaller** than it is.
2) And of course it's no good having equipment that gives really precise readings if it's **inaccurate** — a wrongly calibrated balance that reads 0.2005 g when there's nothing on it, say, is **no good to anyone**.
3) And then there's always room for **human error** — good old-fashioned **silly mistakes**. Human errors can be systematic (e.g. always reading the top of a meniscus rather than the bottom) as well as random.

Practice Questions

Q1 4.6 g of sodium reacts in an experiment. Write down the formula you'd use to calculate how many moles this is.

Q2 Give two systematic problems that can cause the mass of a piece of lithium to be recorded inaccurately.

Q3 Which method should you use to find out how many moles of an alkali you have?

Exam Questions

Q1 Sodium azide, NaN_3, decomposes when heated to form sodium metal, Na, and nitrogen gas, N_2.

a) Write a balanced equation for this reaction. [2 marks]

b) To confirm the equation, 0.325 g of sodium azide was decomposed. The nitrogen gas produced was found to have a volume of 180 cm^3 at room temperature and pressure.

i) How many moles of sodium azide were decomposed? [2 marks]

ii) How many moles of nitrogen gas were formed? [1 mark]

iii) What is the molar ratio of sodium azide to nitrogen? Give your answer as an integer ratio. [2 marks]

Q2 Magnesium reacts with hydrochloric acid according to this equation: $Mg_{(s)} + 2HCl_{(aq)} \rightarrow MgCl_{2\,(aq)} + H_{2\,(g)}$

A student wanted to check that one mole of hydrogen gas is produced for every mole of magnesium that reacts. She put some hydrochloric acid into a conical flask and weighed a piece of magnesium.

a) Describe how she could collect the hydrogen gas produced and measure its volume. [2 marks]

b) Give two reasons why the result of this experiment may not be accurate. [2 marks]

Yup, that's an equation all right...

Who'd have thought it — all that hassle just to check that a reaction happens the way you think it does. However, a bit more practice with those calculations never did anyone any harm. And that stuff about errors is really important — you need to understand the different types of error, and be able to give examples of where they happen in real experiments.

Making Salts

You need to know how to prepare a salt and calculate the percentage yield. And this includes double salts. Fancy.

Salts Can Be Hydrated

1) All solid salts consist of a **lattice** of positive and negative ions. In some salts, **water molecules** are incorporated in the lattice too (it's called water of crystallisation).
2) A solid salt containing water of crystallisation is **hydrated**. The formula of a hydrated salt shows **how many water molecules** are incorporated in the lattice.
3) For example, **hydrated copper sulfate** has **five** moles of water for every mole of the salt. So its formula is $\mathbf{CuSO_4.5H_2O}$.

Notice that there's a dot between $CuSO_4$ and $5H_2O$.

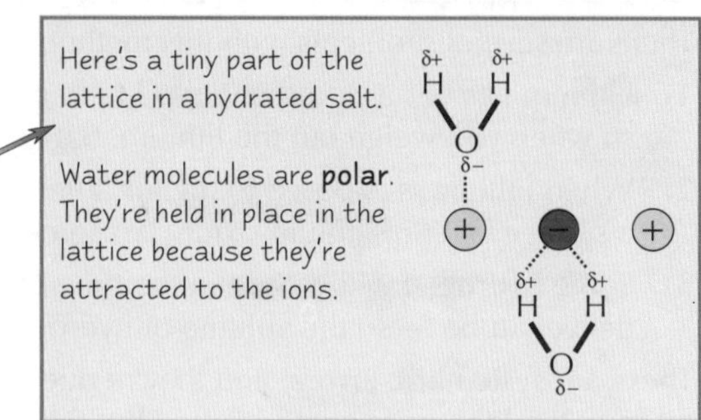

A Double Salt Contains Two Cations or Two Anions

1) As I'm sure you know, you can get a **salt** by reacting an acid and a base. E.g. $NaOH_{(aq)} + HCl_{(aq)} \rightarrow NaCl_{(aq)} + H_2O_{(l)}$
2) Well, if you mix solutions of two different salts and crystallise them, you'll end up with a **double salt** — a salt with two different anions or cations.
3) For example, if you mix and crystallise ammonium sulfate solution and iron(II) sulfate solution, you'll get **ammonium iron(II) sulfate**, $(NH_4)_2Fe(SO_4)_2$. This is a double salt because it's got two cations.
4) But you can also prepare a double salt from raw ingredients, rather than from two salt solutions...

!!!

The double assault was going to plan so far. (Ah, the old classics are the best.)

Preparing a Salt — Mix the Right Stuff, Crystallise, Filter

Here's how to prepare **hydrated ammonium iron(II) sulfate**, $(NH_4)_2Fe(SO_4)_2.6H_2O$ from **iron, ammonia** and **sulfuric acid**.

Here's the balanced equation. It looks worse than it is, I promise.

$$\mathbf{Fe_{(s)} + 2NH_{3\,(aq)} + 2H_2SO_{4\,(aq)} + 6H_2O_{(l)} \rightarrow (NH_4)_2Fe(SO_4)_2.6H_2O_{(s)} + H_{2\,(g)}}$$

1) Add a known mass of **iron filings** to an excess of **warm sulfuric acid** and stir until they've all reacted. You've now got **iron(II) sulfate solution**.
2) Add **just enough** ammonia solution to **react completely** with the iron. (From the equation, you need twice as many moles of ammonia as of iron — you need to work out what amounts to use.)
3) Leave the solution to evaporate — **blue-green crystals** of the salt will form. Some solution will remain (because you started with an excess of acid).
4) Collect the crystals by **filtering**, then wash them using distilled water.
5) To **dry** the crystals, press them between two pieces of filter paper to absorb as much water as possible.

solution/crystal mixture
funnel
filter paper
solution
conical flask

Making Salts

Percentage Yield Is Never 100%

1) The **theoretical yield** is the **mass of product** that **should** be formed. It assumes **no** chemicals are '**lost**' in the process.
2) You can use the **masses of reactants** and a **balanced equation** to calculate the theoretical yield for a reaction.
3) For example, here's how to calculate the theoretical yield of **hydrated ammonium iron(II) sulfate**. Say you react **1.40 g** of **iron filings**:

$$Fe_{(s)} + 2NH_{3\,(aq)} + 2H_2SO_{4\,(aq)} + 6H_2O_{(l)} \rightarrow (NH_4)_2Fe(SO_4)_2.6H_2O_{(s)} + H_{2\,(g)}$$

- Number of moles of **iron** (A_r = 56) reacted = mass ÷ molar mass = 1.40 ÷ 56 = **0.025 moles.**
  From the equation, 'moles of iron : moles of ammonium iron(II) sulfate' is 1 : 1, so 0.025 moles of product should form.
- Molar mass of $(NH_4)_2Fe(SO_4)_2.6H_2O_{(s)}$ = 392 g mol^{-1}, so **theoretical yield** = 0.025 × 392 = **9.8 g.**

4) For any reaction, the **actual** mass of product (the **actual yield**) will always be **less** than the theoretical yield. There are many reasons for this. For example, sometimes not all the 'starting' chemicals react fully. And some chemicals are always 'lost', e.g. some solution gets left on filter paper, or is lost during transfers between containers.
5) So, in the ammonium iron(II) sulfate experiment, the theoretical yield was 9.8 g... but you won't actually **get** 9.8 g.
6) In this case, to find the **actual yield** you just **weigh the crystals**. Then you can work out the **percentage yield**.

$$\text{Percentage yield} = \frac{\text{Actual Yield}}{\text{Theoretical Yield}} \times 100\%$$

So if the actual yield of hydrated ammonium iron(II) sulfate crystals was **5.2 g**:

Percentage yield = (5.2 ÷ 9.8) × 100% = 53%

Practice Questions

Q1 What is a double salt?

Q2 What is the theoretical yield?

Q3 Why is it never possible to prepare a salt with a 100% yield?

Q4 Write down the formula for calculating percentage yield.

Exam Question

Q1 Copper(II) sulfate pentahydrate, $CuSO_4.5H_2O$, was prepared by adding excess copper(II) oxide to 50 cm^3 of hot 0.2 mol dm^{-3} sulfuric acid. The equation for the reaction is:

$$CuO_{(s)} + H_2SO_{4\,(aq)} + 4H_2O_{(l)} \rightarrow CuSO_4.5H_2O_{(aq)}$$

The solution was filtered to remove any unreacted solid and allowed to evaporate until crystals of product formed. The crystals were collected and dried with filter paper. The dry crystals had a mass of 1.964 g.

a) How many moles of sulfuric acid were used? [1 mark]

b) What was the percentage yield for the reaction? [4 marks]

Percentage Revision Yield = Pages Learnt ÷ Pages Read × 100%

Ah, salts. Back in the day, there was only one kind of salt you needed to know about and it usually went on your fish and chips. Alas, those innocent days are gone for ever. Now you need to know about hydrated double salts as well as your bog-standard anhydrous salts — how to write their formulas, how to make dry samples and how to do pesky sums.

Atom Economy and Percentage Yield

How to make a subject like chemistry even more exciting — introduce the word 'economy'...

Atom Economy is a Measure of the **Efficiency** of a Reaction

1) The **efficiency** of a reaction is often measured by the **percentage yield** (see the previous page). This tells you how wasteful the **process** is — it's based on how much of the product is lost because of things like reactions not completing or losses during collection and purification.
2) But percentage yield doesn't measure how wasteful the **reaction** itself is. A reaction that has a 100% yield could still be very wasteful if a lot of the atoms from the **reactants** wind up in **by-products** rather than the **desired product**.
3) **Atom economy** is a measure of the proportion of reactant **atoms** that become part of the desired product (rather than by-products) in the **balanced** chemical equation. It's calculated using this formula:

$$\% \text{ atom economy} = \frac{\text{molecular mass of desired product}}{\text{sum of molecular masses of all products}} \times 100$$

Addition Reactions have a **100%** Atom Economy

1) In an **addition reaction** the reactants **combine** to form a **single product**. The atom economy for addition reactions is **always 100%** since no atoms are wasted.
2) For example, ethene (C_2H_4) and hydrogen react to form ethane (C_2H_6) in an addition reaction:

$$C_2H_4 + H_2 \rightarrow C_2H_6$$

The **only product** is ethane — the desired product. So **none** of the reactant atoms are wasted.

Substitution Reactions have a **Lower Atom Economy** than Addition Reactions

1) A **substitution reaction** is one where some atoms from one reactant are **swapped** with atoms from another reactant. This type of reaction **always** results in **at least two products** — the desired product and at least one by-product.
2) An example is the reaction of bromomethane (CH_3Br) with sodium hydroxide (NaOH) to make methanol (CH_3OH):

$$CH_3Br + NaOH \rightarrow CH_3OH + NaBr$$

Here the Br atoms have **swapped places** with the OH groups. This is **more wasteful** than an addition reaction because the Na and Br atoms are not part of the desired product.
3) The **atom economy** for this reaction is:

Always make sure you're using a balanced equation.

$$\% \text{ atom economy} = \frac{\text{molecular mass of desired product}}{\text{sum of molecular masses of all products}} \times 100$$

$$= \frac{M_r(CH_3OH)}{M_r(CH_3OH) + M_r(NaBr)} \times 100$$

$$= \frac{(12+(3\times1)+16+1)}{(12+(3\times1)+16+1)+(23+80)} \times 100 = \frac{32}{32+103} \times 100 = \mathbf{23.7\%}$$

4) Here's another example:

Example: Aluminium oxide is formed by heating aluminium hydroxide until it decomposes. Calculate the atom economy of the reaction. $2Al(OH)_3 \rightarrow Al_2O_3 + 3H_2O$

$$\% \text{ atom economy} = \frac{\text{molecular mass of desired product}}{\text{sum of molecular masses of all products}} \times 100$$

$$= \frac{M_r(Al_2O_3)}{M_r(Al_2O_3) + M_r(3H_2O)} \times 100$$

$$= \frac{(27\times2)+(16\times3)}{[(27\times2)+(16\times3)]+3\times[(1\times2)+16]} \times 100 = \frac{102}{102+54} \times 100 = \mathbf{65.4\%}$$

Atom Economy and Percentage Yield

A Reaction can Have a **High Percentage Yield** and a **Low Atom Economy**

Example: 0.475 g of CH_3Br reacts with excess NaOH in this reaction: $CH_3Br + NaOH \rightarrow CH_3OH + NaBr$
0.153 g of CH_3OH is produced. What is the percentage yield?

Number of moles = mass of substance ÷ molar mass
Moles of CH_3Br = 0.475 ÷ (12 + 3 × 1 + 80) = 0.475 ÷ 95 = **0.005 moles**
The reactant : product ratio is 1 : 1, so the maximum number of moles of CH_3OH is **0.005**.
Theoretical yield = 0.005 × $M_r(CH_3OH)$ = 0.005 × (12 + (3 × 1) + 16 + 1) = 0.005 × 32 = **0.160 g**

$$\text{percentage yield} = \frac{\text{actual yield}}{\text{theoretical yield}} \times 100 = \frac{0.153}{0.160} \times 100 = \mathbf{95.6\%}$$

This reaction has a **very high percentage yield**, but as we've already seen, the **atom economy** is **low**.

It is Important to Develop Reactions with **High Atom Economies**

1) Companies in the chemical industry will often choose to use reactions with high atom economies. Keeping atom economy as high as possible has **environmental** and **economic benefits**.
2) A **low atom economy** means there's lots of **waste** produced, which has to go somewhere. It costs money to **separate** the desired product from the waste products and more money to dispose of the waste products **safely** so they don't harm the environment. (Finding uses for the by-products helps against this).
3) If a large proportion of the mass of the reactants ends up as waste rather than ending up as useful products, the reactants are being used **inefficiently**. This is costly to the company (who have to buy a large mass of reactant chemicals to make the product). It also lowers the **sustainability** of the process (see page 110) — many raw materials are in limited supply, so it makes sense to use them efficiently so they last as long as possible.

Practice Questions

Q1 How many products are there in an addition reaction?

Q2 Does the percentage yield for a reaction always have the same value as the percentage atom economy?

Q3 Why do reactions with high atom economy save chemical companies money and cause less environmental impact?

Exam Questions

Q1 Reactions 1 and 2 below show two possible ways of preparing the compound chloroethane (C_2H_5Cl):

1 $C_2H_5OH + PCl_5 \rightarrow C_2H_5Cl + POCl_3 + HCl$
2 $C_2H_4 + HCl \rightarrow C_2H_5Cl$

a) Which of these is an addition reaction? [1 mark]

b) Calculate the atom economy for reaction 1. [3 marks]

c) Reaction 2 has an atom economy of 100%. Explain why this is in terms of the products of the reaction. [1 mark]

Q2 Phosphorus trichloride (PCl_3) reacts with chlorine to give phosphorus pentachloride (PCl_5):

$$PCl_3 + Cl_2 \rightleftharpoons PCl_5$$

a) If 0.275 g of PCl_3 reacts with 0.142 g of chlorine, what is the theoretical yield of PCl_5? [2 marks]

b) When this reaction is performed 0.198 g of PCl_5 is collected. Calculate the percentage yield. [1 mark]

c) Changing conditions such as temperature and pressure will alter the percentage yield of this reaction.
Will changing these conditions affect the atom economy? Explain your answer. [2 marks]

I knew a Tommy Conomy once... strange bloke...

These pages shouldn't be too much trouble — you've survived worse already. Make sure you get plenty of practice using the atom economy formula. And don't get mixed up between percentage yield (which is to do with the process) and atom economy (which is to do with the reaction). Remember — just because one's high, it doesn't mean the other is too.

Enthalpy Changes

A whole new section to enjoy — but don't forget, Big Brother is watching...

Chemical Reactions Often Have Enthalpy Changes

When chemical reactions happen, some bonds are **broken** and some bonds are **made**. More often than not, this'll cause a **change in energy**. The souped-up chemistry term for this is **enthalpy change** —

> **Enthalpy change**, ΔH (delta H), is the heat energy transferred in a reaction at **constant pressure**. The units of ΔH are **kJ mol^{-1}**.

You write $\Delta H^{\ominus}$ to show that the elements were in their **standard states** and that the measurements were made under **standard conditions**. Standard conditions are **100 kPa (about 1 atm) pressure** and a temperature of **298 K (25 °C)**. The next page explains why this is necessary.

Reactions can be either Exothermic or Endothermic

> **Exothermic** reactions **give out** energy. ΔH is **negative.**

In exothermic reactions, the temperature of the surroundings often goes **up**. That's how **hand-warmers** work — breaking the seal lets two substances mix, they react and the pack **gives out heat** to your hands.

Oxidation is exothermic. Here are two examples —

- The **combustion** of a fuel like methane ⟹ $CH_{4(g)} + 2O_{2(g)} \rightarrow CO_{2(g)} + 2H_2O_{(l)}$ $\Delta H_c^{\ominus} = -890$ kJ mol^{-1} **exothermic**
- The oxidation of **carbohydrates**, such as glucose, $C_6H_{12}O_6$, in respiration.

> **Endothermic** reactions **absorb** energy. ΔH is **positive.**

In these reactions, the temperature of the surroundings often **falls**. For example, **cold packs** for treating sports injuries work by gaining heat **from** your body.

The **thermal decomposition** of calcium carbonate is endothermic.

$$CaCO_{3(s)} \rightarrow CaO_{(s)} + CO_{2(g)} \quad \Delta H_r^{\ominus} = +178 \text{ kJ mol}^{-1} \text{ endothermic}$$

The main reactions of **photosynthesis** are also endothermic — sunlight supplies the energy.

Enthalpy Profile Diagrams Show Energy Change in Reactions

1) **Enthalpy Profile Diagrams** show you how the enthalpy (energy) changes during reactions.
2) The **activation energy**, E_a, is the minimum amount of energy needed to begin breaking reactant bonds and start a chemical reaction.

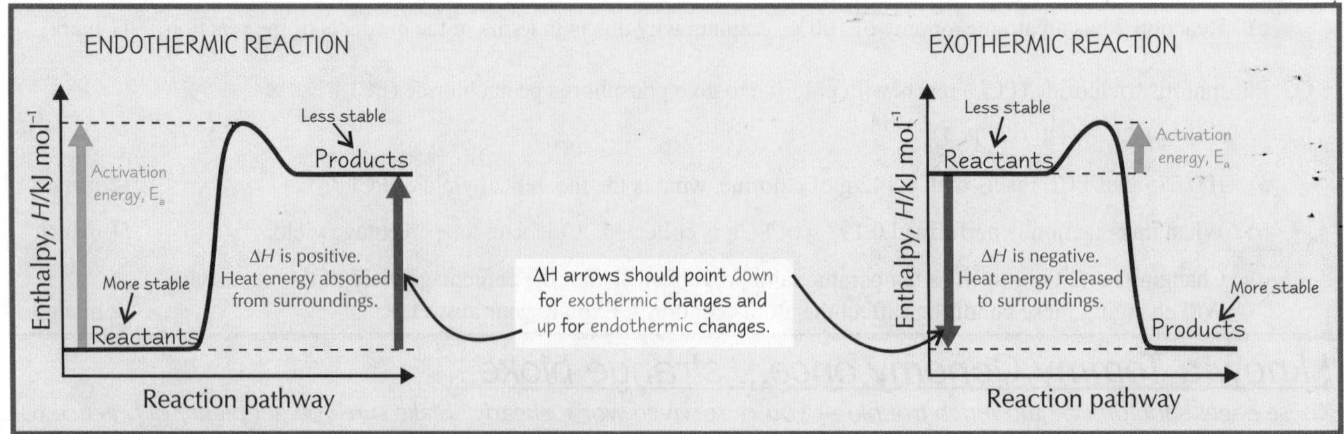

3) The **less enthalpy** a substance has, the **more stable** it is.

Enthalpy Changes

You Need to Specify the **Conditions** for **Enthalpy Changes**

1) You can't directly measure the **actual** enthalpy of a system. In practice, that doesn't matter, because it's only ever **enthalpy change** that matters.
2) You can find enthalpy changes either by **experiment** or in **textbooks**. Enthalpy changes you find in textbooks are usually **standard** enthalpy changes — enthalpy changes under **standard conditions** (**298 K** and **100 kPa**).
3) This is important because changes in enthalpy are affected by **temperature** and **pressure** — using standard conditions means that everyone can know **exactly** what the enthalpy change is describing.

There are Different Types of ΔH Depending On the **Reaction**

1) **Standard enthalpy change of reaction**, $\Delta H_r^\ominus$, is the enthalpy change when the reaction occurs in the **molar quantities** shown in the **chemical equation**, under standard conditions in their standard states.
2) **Standard enthalpy change of formation**, $\Delta H_f^\ominus$, is the enthalpy change when **1 mole** of a **compound** is formed from its **elements** in their standard states under standard conditions, e.g. $2C_{(s)} + 3H_{2(g)} + \frac{1}{2}O_{2(g)} \longrightarrow C_2H_5OH_{(l)}$
3) **Standard enthalpy change of combustion**, $\Delta H_c^\ominus$, is the enthalpy change when **1 mole** of a substance is completely **burned in oxygen** under standard conditions.
4) **Standard enthalpy of neutralisation**, $\Delta H_{neut}^\ominus$, is the enthalpy change when **1 mole** of **water** is formed from the neutralisation of **hydrogen ions** by **hydroxide ions** under standard conditions, e.g. $H^+_{(aq)} + OH^-_{(aq)} \longrightarrow H_2O_{(l)}$
5) **Standard enthalpy change of atomisation**, $\Delta H_{at}^\ominus$, is the enthalpy change when **1 mole** of **gaseous atoms** is formed from the element in its **standard state**, e.g. $\frac{1}{2}Cl_{2(g)} \longrightarrow Cl_{(g)}$

Practice Questions

Q1 Explain the terms exothermic and endothermic, giving an example in each case.

Q2 Draw and label enthalpy profile diagrams for an exothermic and an endothermic reaction.

Q3 Define standard enthalpy of formation and standard enthalpy of combustion.

Q4 Why are the enthalpy changes given in data books called 'standard' enthalpy changes?

Exam Questions

Q1 Hydrogen peroxide, H_2O_2, can decompose into water and oxygen. $2H_2O_{2(l)} \longrightarrow 2H_2O_{(l)} + O_{2(g)}$ $\Delta H_r^\ominus = -98 \text{ kJ mol}^{-1}$
Draw an enthalpy profile diagram for this reaction. Mark on the activation energy and ΔH. [3 marks]

Q2 Methanol, CH_3OH, when blended with petrol, can be used as a fuel. $\Delta H_c^\ominus[CH_3OH] = -726 \text{ kJ mol}^{-1}$.

a) Write an equation, including state symbols, for the standard enthalpy change of combustion of methanol. [2 marks]

b) Write an equation, including state symbols, for the standard enthalpy change of formation of methanol. [2 marks]

c) Liquid petroleum gas is a fuel that contains propane, C_3H_8.
Explain why the following equation does not represent a standard enthalpy change of combustion. [1 mark]

$$2C_3H_{8(g)} + 10O_{2(g)} \longrightarrow 8H_2O_{(l)} + 6CO_{2(g)} \qquad \Delta H_r = -4113 \text{ kJ mol}^{-1}$$

It's getting hot in here, so take off all your bonds...

What a lotta definitions. And you need to know them all. If you're going to bother learning them, you might as well do it properly and learn all the pernickety details. They probably seem about as useful as a dead fly in your custard right now, but all will be revealed over the next few pages. Learn them now, so you've got a bit of a head start.

Finding Enthalpy Changes

Now you know what enthalpy changes are, here's how to calculate them...

You can find out Enthalpy Changes *using* Calorimetry

In **calorimetry** you find how much heat is given out by a reaction by measuring the **temperature change** of some water.

1) To find the enthalpy of **combustion** of a **flammable liquid**, you burn it — using apparatus like this...
2) As the fuel burns, it heats the water. You can work out the **heat absorbed** by the water if you know the **mass of water**, the **temperature change of the water** (ΔT), and the **specific heat capacity of water** (= 4.18 J g^{-1} K^{-1}) — see below for the details.
3) Ideally, all the heat given out by the fuel as it burns would be **absorbed** by the water — allowing you to work out the enthalpy change of combustion (see below). In practice, you **always** lose some heat (as you heat the apparatus and the surroundings).

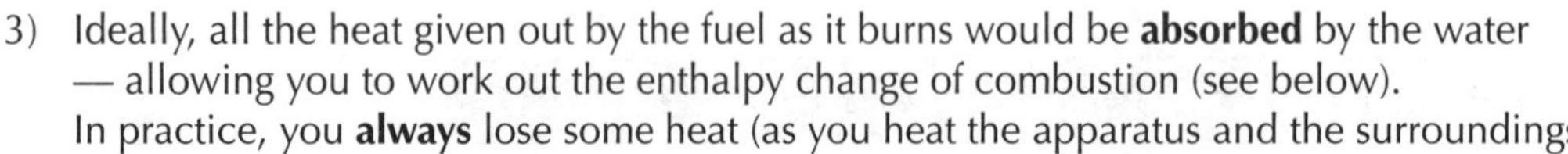

Calorimetry can also be used to calculate the enthalpy change for a reaction that happens **in solution**, such as **neutralisation** or **displacement**.

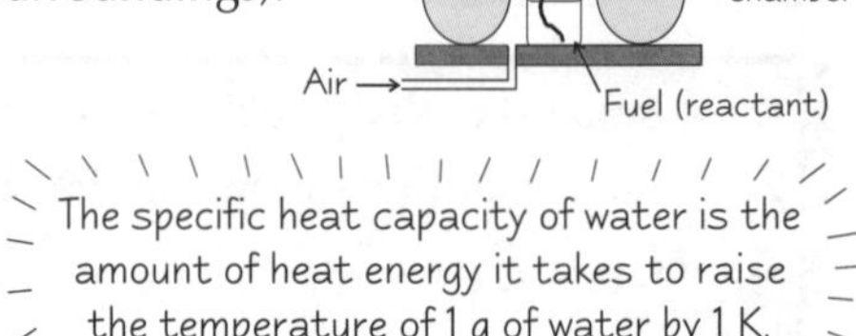
The specific heat capacity of water is the amount of heat energy it takes to raise the temperature of 1 g of water by 1 K.

1) To find the enthalpy change in a neutralisation reaction, add a **known volume** of acid to an **insulated container** and measure the **temperature**.
2) Then add a **known volume** of alkali, and record the **temperature rise**. (Stir the solution to make sure the solution is evenly heated.)
3) You can work out the heat needed to **raise the temperature** of the solution formed using the formula below — this **equals** the **heat given out** by the **reaction**.
4) You can usually assume that all solutions (reactants and product) have the **same density as water**. This means you can use **volume** (rather than mass) in your calculations (as 1 cm^3 of water has a mass of 1 g).

Calculate Enthalpy Changes *Using the* Equation $q = mc\Delta T$

It seems there's a snazzy equation for everything these days, and enthalpy change is no exception:

$q = mc\Delta T$ where, q = heat lost or gained (in joules). This is the same as the enthalpy change if the pressure is constant.

m = mass of water in the calorimeter, or solution in the polystyrene beaker (in grams)

c = specific heat capacity of water (4.18 J $g^{-1}K^{-1}$)

ΔT = the change in temperature of the water or solution

Example:

In a laboratory experiment, 1.16 g of an organic liquid fuel was completely burned in oxygen.
The heat formed during this combustion raised the temperature of 100 g of water from 295.3 K to 357.8 K.
Calculate the standard enthalpy of combustion, ΔH_c, of the fuel. Its M_r is 58.

Remember — m is the mass of water, NOT the mass of fuel.

1. First off, you need to calculate the **amount of heat** given out by the fuel using $q = mc\Delta T$.

$q = mc\Delta T$

$q = 100 \times 4.18 \times (357.8 - 295.3) = 26\,125 \text{ J} = 26.125 \text{ kJ}$ ⟸ Change the amount of heat from J to kJ.

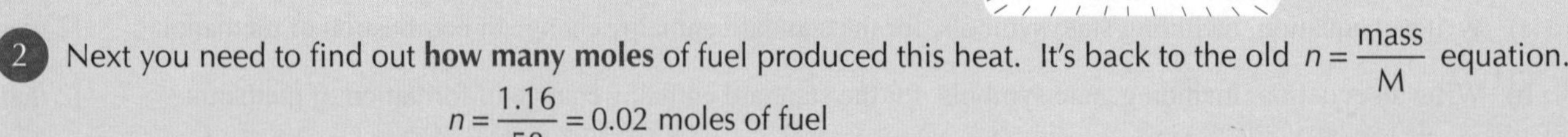

2. Next you need to find out **how many moles** of fuel produced this heat. It's back to the old $n = \frac{\text{mass}}{M}$ equation.

$$n = \frac{1.16}{58} = 0.02 \text{ moles of fuel}$$

3. The standard enthalpy of combustion involves 1 mole of fuel.

It's negative because combustion is an exothermic reaction.

So, the heat produced by 1 mole of fuel $= \frac{-26.125}{0.02}$

$\approx$ **-1306 kJ mol^{-1}**. This is the standard enthalpy change of combustion.

The actual ΔH_c of this compound is -1615 kJ mol^{-1} — loads of heat has been **lost** and not measured. E.g. it's likely a fair bit would escape through the **copper calorimeter** and also the fuel might not **combust completely**.

Finding Enthalpy Changes

Experimental Results **Always** Include **Errors**

Types of error were also mentioned on page 13.

You need know about **errors** in experiments — there are **two kinds**, and you always get **both** when you do an experiment.

1) **Systematic errors** are repeated **every time** you carry out the experiment, and always affect your result in the **same way** (e.g. they always make your answer bigger than it should be, or always make it smaller than it should be).
   They're due to the **experimental set-up**, or **limitations of the equipment**.
   For example, a balance that always reads 0.2 g less than the true value will result in systematic errors.
2) Here are some examples of **systematic errors** in calorimetry experiments...

Experimental problems with calorimetry generally...

- Some heat will be **absorbed** by the **container**, rather than going towards heating up the **water**.
- Some heat is always **lost to the surroundings** during the experiment (however well you **insulate** the container).

Experimental problems with flammable-liquid calorimetry...

- Some combustion may be **incomplete** — which will mean **less energy** will be given out.
- Some of the flammable liquid may escape by **evaporation** (they're usually quite **volatile**).

3) **Random errors** are... random — there's no pattern to them. And they **always** happen.
   The best way to deal with these is to **repeat** your experiment, and take the **average** of all your readings (see below).

Accuracy and ***Reliability*** are ***NOT*** the Same

Ideally you want both of these.

1) **Accuracy** and **reliability** are not the same thing.
   Accuracy means 'how close to the true value' your results are.
   Reliability means 'how reproducible' your results are.
2) **Repeating** an experiment shows whether your results are **reliable**.
   If you repeat an experiment and find the average (the mean), the effect of **random errors** is reduced — **positive** random errors and **negative** random errors should mostly **cancel out**. The more times you repeat an experiment, the more reliable your mean is.
3) But reliable results **aren't** necessarily more **accurate**.
   Repeating an experiment doesn't do anything to eliminate **systematic** errors.

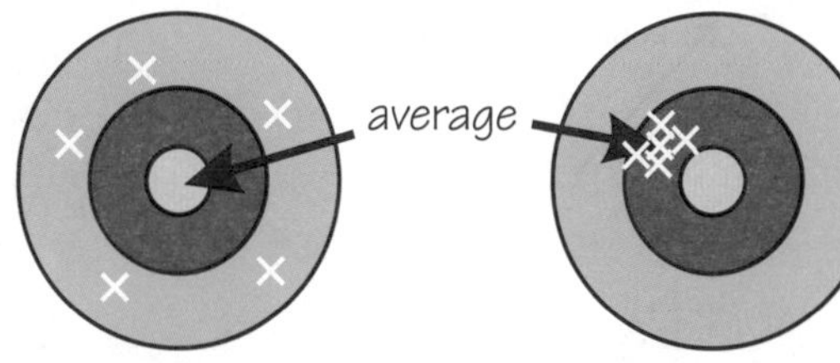

Accurate but not reliable — large random errors, but centred around the bullseye. The average is spot on.

Reliable but not accurate — large systematic errors, but very precise.

Practice Questions

Q1 Briefly describe an experiment that could be carried out to find the enthalpy change of a reaction.

Q2 Why is the enthalpy change determined in a laboratory likely to be lower than the value shown in a data book?

Q3 What equation is used to calculate the heat change in a chemical reaction?

Exam Questions

Q1 The initial temperature of 25 cm^3 of 1.0 mol dm^{-3} hydrochloric acid in a polystyrene cup was measured as 19 °C.
This acid was exactly neutralised by 25 cm^3 of 1.0 mol dm^{-3} sodium hydroxide solution, also at 19 °C when added.
The maximum temperature of the resulting solution was measured as 25.5 °C.

Calculate the molar enthalpy change of this reaction. (Assume that the neutral solution formed has a specific heat capacity of 4.18 J K^{-1} g^{-1}, and a density of 1.0 g cm^{-3}.) [7 marks]

Q2 A 50 cm^3 sample of 0.2 M copper(II) sulfate solution placed in a polystyrene beaker gave a temperature increase of 2.6 K when excess zinc powder was added and stirred. (Ignore the increase in volume due to the zinc.)
Calculate the enthalpy change when 1 mole of zinc reacts. Assume the solution's specific heat capacity is 4.18 J $g^{-1}K^{-1}$.
The equation for the reaction is: $Zn_{(s)} + CuSO_{4(aq)} \rightarrow Cu_{(s)} + ZnSO_{4(aq)}$ [6 marks]

Experimental errors can make life such a trial...

Errors always happen — that's a fact of life. But that doesn't mean you can just turn your mind off to them. Sometimes you might be able to see what your systematic errors are, and do something about them. As for random errors... well, not being slapdash about things, plus repeating your experiments, can help reduce those. It's all quite "How Science Works-y", this.

Using Hess's Law

Sometimes you can't work out an enthalpy change by measuring a single temperature change. But there's still a way.

Hess's Law — the Total Enthalpy Change is Independent of the Route Taken

Hess's Law says that:

The **total enthalpy change** of a reaction is always **the same**, no matter **which route** is taken.

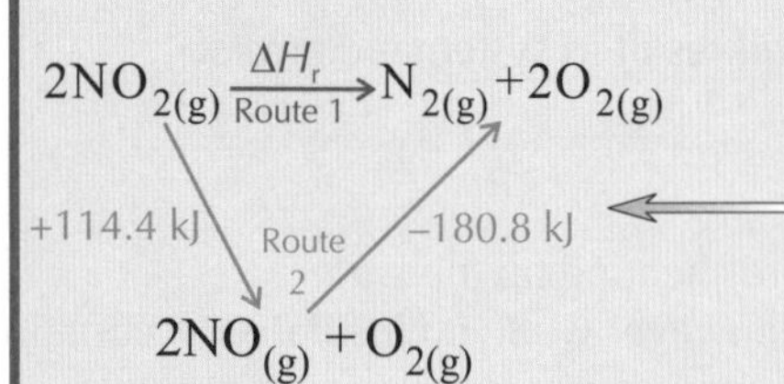

This law is handy for working out enthalpy changes that you **can't find directly** by doing an experiment.

Here's an example:
The **total enthalpy change** for route 1 is the **same as for route 2**.
So, ΔH_r = +114.4 + (−180.8) = −66.4 kJ mol^{-1}.

Enthalpy Changes Can be Worked Out From Enthalpies of Formation...

Enthalpy changes of formation are useful for calculating enthalpy changes you can't find directly.
You need to know $\Delta H_f^\ominus$ for **all** the reactants and products that are **compounds**.

The value of $\Delta H_f^\ominus$ for elements is **zero** — the element's being formed from the element, so there's no change.

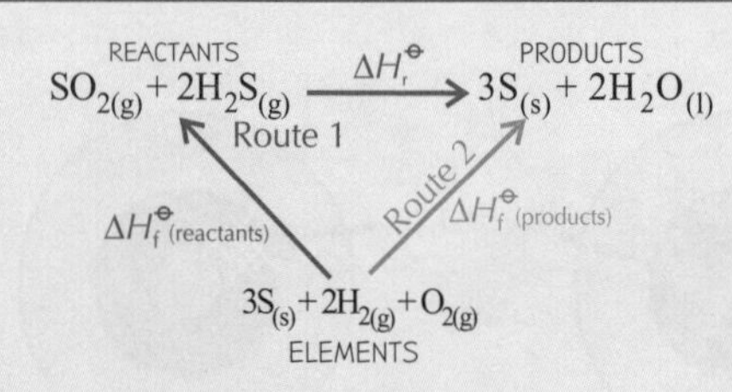

$\Delta H_f^\ominus[SO_{2(g)}]$ = −297 kJ mol^{-1}

$\Delta H_f^\ominus[H_2S_{(g)}]$ = −20.2 kJ mol^{-1}

$\Delta H_f^\ominus[H_2O_{(l)}]$ = −286 kJ mol^{-1}

Here's how to calculate $\Delta H_r^\ominus$ for the reaction shown...

Using **Hess's Law**: Route 1 = Route 2

$\Delta H_r^\ominus$ + the sum of $\Delta H_f^\ominus$ (reactants) = the sum of $\Delta H_f^\ominus$ (products)

So, **$\Delta H_r^\ominus$ = the sum of $\Delta H_f^\ominus$ (products) − the sum of $\Delta H_f^\ominus$ (reactants)**

To find $\Delta H_r^\ominus$ of this reaction: $SO_{2(g)} + 2H_2S_{(g)} \rightarrow 3S_{(s)} + 2H_2O_{(l)}$

Just plug the numbers into the equation above:

$\Delta H_r^\ominus$ = [0 + (−286 × 2)] − [−297 + (−20.2 × 2)] = **−234.6 kJ mol^{-1}**

$\Delta H_f^\ominus$ of sulfur is zero — it's an element.

There's 2 moles of H_2O and 2 moles of H_2S.

It **always** works, no matter how complicated the reaction — e.g. $2NH_4NO_{3\,(s)} + C_{(s)} \rightarrow 2N_{2\,(g)} + CO_{2\,(g)} + 4H_2O_{(l)}$

Using Hess's Law: Route 1 = Route 2

$\Delta H_f^\ominus$[reactants] + $\Delta H_r^\ominus$ = $\Delta H_f^\ominus$[products]

2 × −365 + 0 + $\Delta H_r^\ominus$ = 0 + −394 + (4 × −286)

$\Delta H_r^\ominus$ = −394 + (−1144) − (−730)
= **−808 kJ mol^{-1}**.

Remember... $\Delta H_f^\ominus$ for <u>any</u> element is zero.

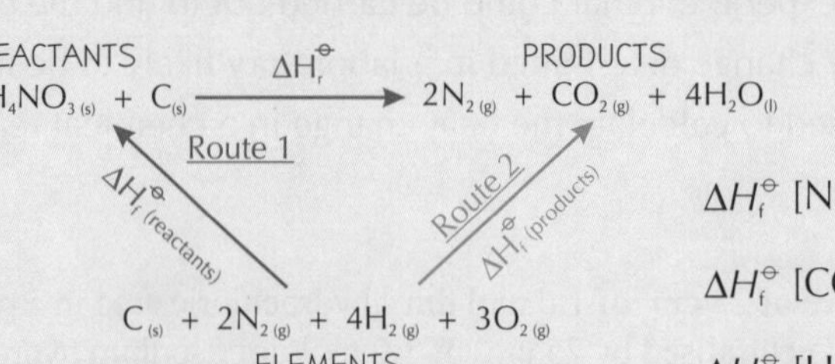

$\Delta H_f^\ominus$ $[NH_4NO_3]_{(s)}$ = −365 kJ mol^{-1}

$\Delta H_f^\ominus$ $[CO_2]_{(g)}$ = −394 kJ mol^{-1}

$\Delta H_f^\ominus$ $[H_2O]_{(l)}$ = −286 kJ mol^{-1}

...And From Enthalpies of Combustion

You can use a similar method to find **enthalpy changes of formation** from enthalpy changes of combustion.

<u>Here's how to calculate $\Delta H_f^\ominus$ of ethanol</u>...

Using Hess's Law: Route 1 = Route 2

$\Delta H_f^\ominus$[ethanol] + $\Delta H_c^\ominus$[ethanol] = 2$\Delta H_c^\ominus$[C] + 3$\Delta H_c^\ominus[H_2]$

$\Delta H_f^\ominus$[ethanol] + (−1367) = (2 × −394) + (3 × −286)

$\Delta H_f^\ominus$[ethanol] = −788 + −858 − (−1367) = **−279 kJ mol^{-1}**.

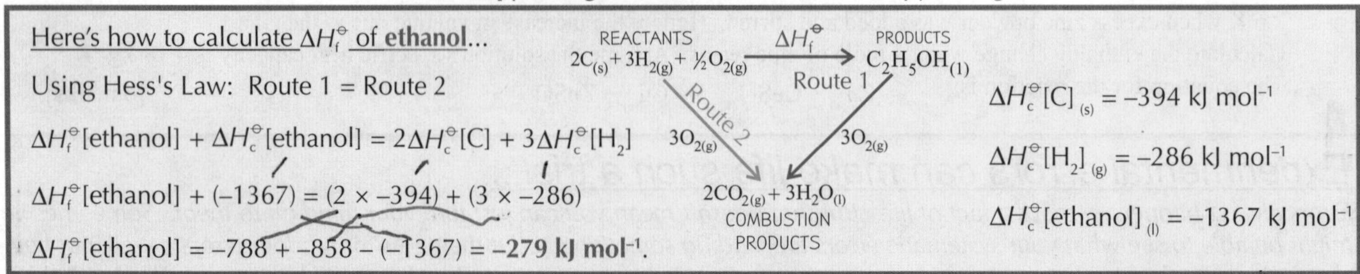

$\Delta H_c^\ominus[C]_{(s)}$ = −394 kJ mol^{-1}

$\Delta H_c^\ominus[H_2]_{(g)}$ = −286 kJ mol^{-1}

$\Delta H_c^\ominus[\text{ethanol}]_{(l)}$ = −1367 kJ mol^{-1}

Using Hess's Law

On p20 you saw how you could find the enthalpy change of a reaction using calorimetry. Sometimes you can **combine** the enthalpy change results from these experiments (neutralisations, for example) to work out an enthalpy change that you **can't find directly**. It's clever stuff... read on.

Hess's Law Lets You Find Enthalpy Changes Indirectly From Experiments

You **can't** find the enthalpy change of the thermal decomposition of calcium carbonate by measuring a temperature change. ⟹ $CaCO_{3(s)} \rightarrow CaO_{(s)} + CO_{2(g)}$ Enthalpy change = ?

(It's an **endothermic reaction**, so you'd expect the temperature to fall. But you need to **heat it up** for the reaction to happen at all).

But you can find it in a more **indirect** way.
The aim is to make one of those **Hess cycles** (the technical name for a "Hess's Law triangle diagram thing").

1. As always, start by drawing the top of the triangle — include your **reactants** and **products**:

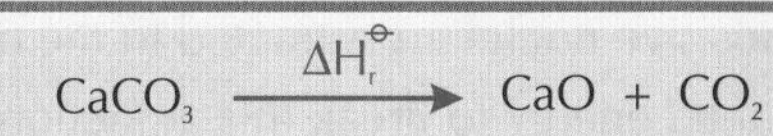

2. Next, you're going to carry out two **neutralisation** reactions involving **hydrochloric acid**, and use the results to complete your Hess cycle. You **can** find the enthalpy changes of these reactions (using calorimetry — see p20). Call them ΔH_1 and ΔH_2.

   Reaction 1: $CaCO_3 + 2HCl \rightarrow CaCl_2 + CO_2 + H_2O$ ΔH_1
   Reaction 2: $CaO + 2HCl \rightarrow CaCl_2 + H_2O$ ΔH_2

3. Now you can build the other two sides of your Hess cycle. Add **2 moles of HCl** to both sides of your triangle's top (representing the 2 moles of HCl in the above equations).

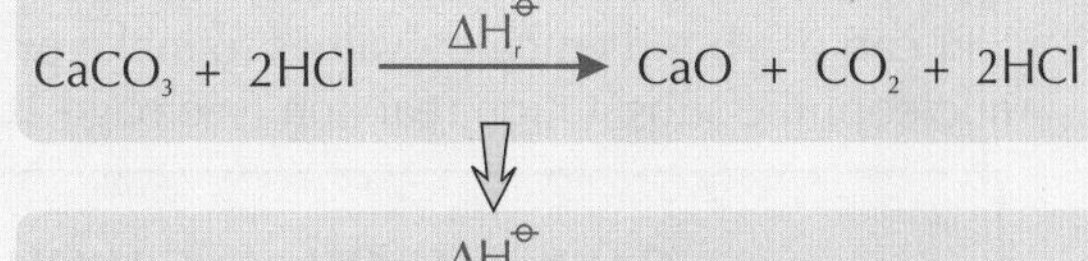

   And add the **products** of the neutralisations to the bottom of the triangle. Notice how all three corners 'balance'.

   $CaCO_3 + 2HCl \xrightarrow{\Delta H_r^\ominus} CaO + 2HCl + CO_2$
   Reaction 1 ΔH_1 ↘ ↙ Reaction 2 ΔH_2
   $CaCl_2 + H_2O + CO_2$

4. Add the enthalpy changes you found to your diagram.
5. And do the maths... the enthalpy change you want to find is just: $\Delta H_1 - \Delta H_2$

Practice Questions

Q1 What does Hess's Law state?

Q2 What is the standard enthalpy change of formation of any element?

Q3 Describe how you can make a Hess cycle to find the standard enthalpy change of a reaction using standard enthalpy changes of formation.

Exam Questions

Q1 Using the facts that the standard enthalpy change of formation of $Al_2O_{3(s)}$ is –1676 kJ mol^{-1} and the standard enthalpy change of formation of $MgO_{(s)}$ is –602 kJ mol^{-1}, calculate the enthalpy change of the following reaction.

$$Al_2O_{3(s)} + 3Mg_{(s)} \rightarrow 2Al_{(s)} + 3MgO_{(s)}$$ [3 marks]

Q2 Calculate the enthalpy change for the reaction below (the fermentation of glucose).

$$C_6H_{12}O_{6\,(s)} \rightarrow 2C_2H_5OH_{(l)} + 2CO_{2\,(g)}$$

Use the following standard enthalpies of combustion in your calculatons:

$\Delta H_c^\ominus$(glucose) = –2820 kJ mol^{-1}, $\Delta H_c^\ominus$(ethanol) = –1367 kJ mol^{-1} [3 marks]

To understand this lot, you're gonna need a bar of chocolate. Or two...

To get your head around those Hess diagrams, you're going to have to do more than skim them. It'll also help if you know the definitions for those standard enthalpy thingumabobs on page 19. If you didn't bother learning them, have a quick flick back and remind yourself about them — especially the standard enthalpy changes of combustion and formation.

Bond Enthalpy

If it weren't for copyright law, there'd be pictures of James Bond in his swimwear on this page.

Reactions are all about Breaking and Making Bonds

When reactions happen, **reactant bonds** are **broken** and **product bonds** are **formed**.

1) You **need** energy to break bonds, so bond breaking is **endothermic** (ΔH is **positive**).
2) Energy is **released** when bonds are formed, so this is **exothermic** (ΔH is **negative**).
3) The **enthalpy change** for a reaction is the **overall effect** of these two changes. If you need **more** energy to **break** bonds than is released when bonds are made, ΔH is **positive**. If it's less, ΔH is negative.

You can only break bonds if you've got enough energy.

You need Energy to Break the Attraction between Atoms or Ions

1) In ionic bonding, **positive** and **negative ions** are attracted to each other. In covalent molecules, the **positive nuclei** are attracted to the **negative** charge of the shared electrons in a covalent bond.
2) You need energy to **break** this attraction — **stronger** bonds take more energy to break. The **amount of energy** you need **per mole** is called the **bond dissociation enthalpy**, or just **bond enthalpy**. (Of course it's got a fancy name — this is chemistry.)
3) Bond enthalpies always involve bond breaking in **gaseous compounds**. This makes comparisons fair.
4) Bond enthalpies influence **how quickly** a reaction will occur. In general, the **smaller** the **bond enthalpies** of the bonds that need to be broken, the faster a reaction will be at room temperature. That's because **less energy** has to be taken in from the surroundings to break the reactant bonds.
5) You can use bond enthalpy values to make **predictions** about rate of reaction. For example, take this nucleophilic substitution reaction (see page 104) between a halogenoalkane and sodium hydroxide solution:

$$C_2H_5X + NaOH \rightarrow C_2H_5OH + NaX$$

← X is a halogen — e.g. Cl, Br or I.

- For this reaction to occur, the carbon–halogen bond (C–X) has to break. So the **weaker** the C–X bond (the **smaller** the bond enthalpy), the **faster** this reaction will be.
- Here are the bond enthalpies of some C–X bonds: C–Cl: 346 kJ mol^{-1}, C–Br: 292 kJ mol^{-1}, C–I: 228 kJ mol^{-1}
- From these C–X bond enthalpy values you can predict that **iodoethane** will react the **fastest** in aqueous alkali, while chloroethane will react most slowly.

Average Bond Enthalpies are Not Exact

1) There isn't just one bond enthalpy value between two particular types of atom. For example, water (H_2O) has **two O–H bonds**. You'd think it'd take the same amount of energy to break them both... but it **doesn't** — because breaking the first bond changes the 'environment' of the remaining bond:

- Breaking the **first** bond, H–OH$_{(g)}$, takes **492** kJ mol^{-1}. This is written ***E*(H–OH) = +492 kJ mol^{-1}**.
- Breaking the **second** bond, H–O$_{(g)}$, needs a bit less energy. *E*(H–O) = **+428** kJ mol^{-1}. (OH is easier to break apart than H_2O because there's extra electron repulsion.)

2) So what you can do is give the **mean** bond enthalpy — just take an average:

   Using the data for water above, this would be $\frac{492+428}{2}$ = **+460 kJ mol^{-1}**.
3) The **data book** says the bond enthalpy for O–H is +463 kJ mol^{-1}. It's a bit different because it's the average for a **much bigger range** of molecules, not just water...
4) ...for example, the O–H bond in an **alcohol** molecule has carbon atoms and more hydrogen atoms nearby — its environment is different from that in a water molecule, so the bond's strength will be slightly different.
5) So when you look up an **mean bond enthalpy**, what you get is:

That's not quite what I meant by 'changing the environment'.

the energy needed to break one mole of bonds in the gas phase, averaged over many different compounds

Breaking bonds is always an endothermic process, so average bond dissociation enthalpies are always **positive**.

Bond Enthalpy

You Can Use **Bond Enthalpies** in **Hess's Law** Cycles

There's not much you can't do with Hess's Law.* For example, here's how to calculate enthalpy changes using bond enthalpies:

Example: Calculate the enthalpy of formation of methane, $\Delta H_f^\ominus$ $CH_{4(g)}$, using the data below in a Hess's law energy cycle.

$\Delta H_{at}^\ominus$ $C_{(graphite)}$ = +715 kJ mol^{-1}, $\Delta H_{at}^\ominus$ H = +218 kJ mol^{-1}, E(C–H) = +412 kJ mol^{-1}

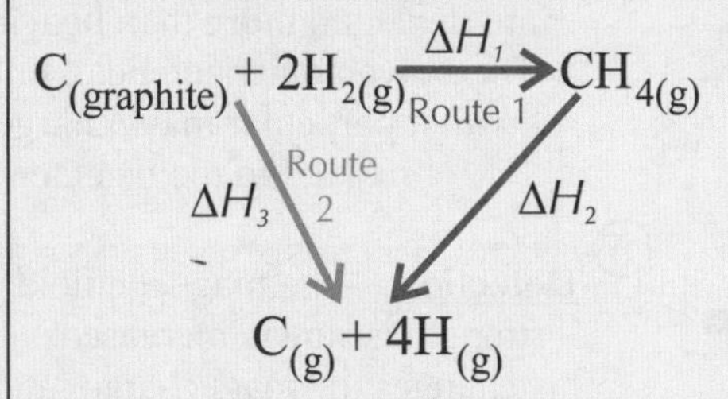

ΔH_1 = enthalpy of formation of methane (what you want to find)

$\Delta H_2 = 4 \times E(C–H) = 4 \times +412 = +1648$ kJ mol^{-1}

$\Delta H_3 = \Delta H_{at}^\ominus C + 4 \times \Delta H_{at}^\ominus H = 715 + 4 \times 218 = +1587$ kJ mol^{-1}

Route 1 = Route 2

$\Delta H_1 + \Delta H_2 = \Delta H_3$

Enthalpy of formation of methane = $\Delta H_1 = \Delta H_3 - \Delta H_2 = 1587 - 1648 =$ **–61 kJ mol^{-1}**

*For those few things, you need duct tape instead.

Practice Questions

Q1 Is energy taken in or released when bonds are broken?

Q2 What state must compounds be in when bond dissociation enthalpies are measured?

Q3 Define average bond enthalpy.

Exam Questions

Q1 a) Construct a Hess's law energy cycle to show the standard enthalpy change of formation of ammonia and the standard enthalpy changes of atomisation of its elements. [4 marks]

b) Use the cycle from part a) and the enthalpy changes given below to calculate the standard enthalpy change for the formation of ammonia.

E(N–H) in ammonia $\Delta H^\ominus$ = +391 kJ mol^{-1}

$½N_{2(g)} \rightarrow N_{(g)}$ $\Delta H^\ominus$ = +473 kJ mol^{-1}

$½H_{2(g)} \rightarrow H_{(g)}$ $\Delta H^\ominus$ = +218 kJ mol^{-1} [4 marks]

c) The data book value for the average bond enthalpy of N–H is +388 kJ mol^{-1}. Why is there a discrepancy between this value and the value used above? [1 mark]

Q2 This table shows some mean bond enthalpy values.

Bond	C–H	C=O	O=O	O–H
Mean Bond Enthalpy (kJ/mol)	435	805	498	464

The complete combustion of methane can be represented by this equation: $CH_{4\,(g)} + 2O_{2\,(g)} \rightarrow CO_{2\,(g)} + 2H_2O_{\,(l)}$

Use the table of bond enthalpies above to suggest which type of reactant bond in this reaction is likely to break first. Explain your answer. [2 marks]

I bonded with my friend. Now we're waiting to be surgically separated...

Reactions are like pulling your Lego spaceship apart and building something new. Sometimes the bits get stuck together and you need to use loads of energy to pull 'em apart. Okay, so energy's not really released when you stick them together, but you can't have everything — and it wasn't that bad an analogy up till now. Ah, well...you best get on and learn this stuff.

Mass Spectrometry

Say you had a mystery ball of gloop — mass spectrometry could tell you what was in the gloop.

Relative Masses can be Measured Using a Mass Spectrometer

You can use a **mass spectrometer** to find out loads of stuff. It can tell you the **relative atomic mass**, **relative molecular mass**, **relative isotopic abundance**, **molecular structure** and your **horoscope** for the next fortnight.

There are **5** things that happen when a sample is squirted into a mass spectrometer.

1. Vaporisation — the sample is turned into **gas (vaporised)** using an electrical heater.
2. Ionisation — the gas particles are bombarded with **high-energy electrons** to ionise them. Electrons are knocked off the particles, leaving **positive ions.**
3. Acceleration — the positive ions are accelerated by an **electric field.**
4. Deflection — The positive ions' paths are altered with a **magnetic field. Lighter ions** have less momentum and are deflected **more** than heavier ions. For a given magnetic field, **only** ions with a particular **mass/charge ratio** make it to the detector.
5. Detection — the magnetic field strength is **slowly increased.** This **changes** the mass/charge ratio of ions that can reach the detector. A **mass spectrum** is produced.

vacuum
heavier ions
high-energy electrons
lighter ions

A Mass Spectrum

The **y-axis** gives the **abundance of ions**, often as a percentage. For an element, the **height** of each peak gives the **relative isotopic abundance**, e.g. 75.5% are the ^{35}Cl isotope.

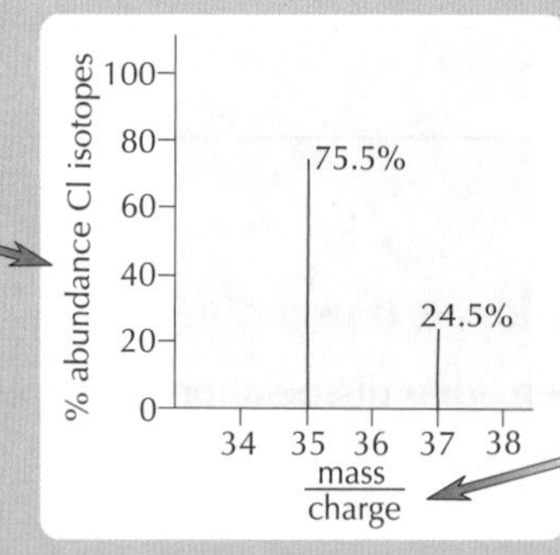

If the sample is an **element**, each line will represent a **different isotope** of the element.

The **x-axis** units are given as a **'mass/charge'** ratio. Since the charge on the ions is mostly **1+**, you can often assume the x-axis is simply the **relative mass**.

A_r and Relative Isotopic Abundance can be Worked Out from a Mass Spectrum

You need to know how to calculate the **relative atomic mass** (A_r) of an element from the **mass spectrum**.

Here's how to calculate A_r for magnesium, using the mass spectrum below —

Step 1: For each peak, read the **% relative isotopic abundance** from the y-axis and the **relative isotopic mass** from the x-axis. **Multiply** them together to get the total mass for each isotope. $79 \times 24 = 1896$; $10 \times 25 = 250$; $11 \times 26 = 286$

Step 2: **Add** up these totals. $1896 + 250 + 286 = 2432$

Step 3: **Divide by 100** (since percentages were used). $A_r(\text{Mg}) = \frac{2432}{100} = 24.32 \approx \mathbf{24.3}$

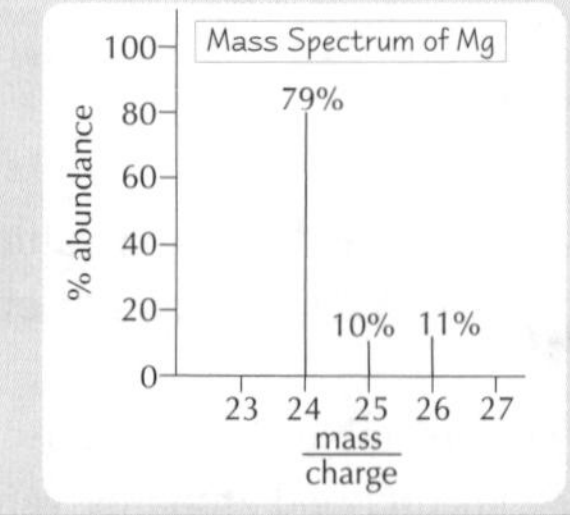

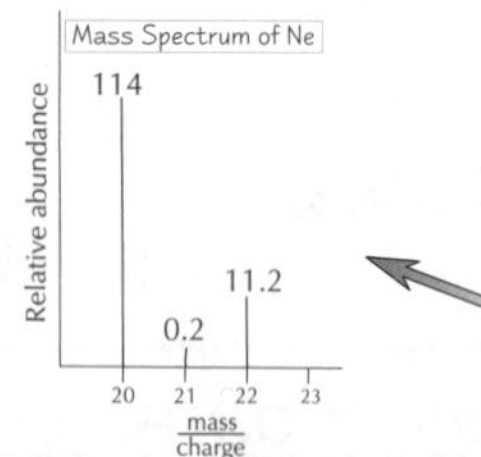

If the relative abundance is **not** given as a percentage, the total abundance may not add up to 100. In this case, don't panic. Just do steps 1 and 2 as above, but then divide by the **total relative abundance** instead of 100 — like this:

$$A_r(\text{Ne}) = \frac{(114 \times 20) + (0.2 \times 21) + (11.2 \times 22)}{114 + 0.2 + 11.2} \approx 20.18$$

Mass spectrometry is a good way to identify elements and molecules (it's kind of like fingerprinting).

Mass Spectrometry

Mass Spectrometry can be used to Find out M_r

You can also get a mass spectrum for a **molecular sample**, such as ethanol (CH_3CH_2OH).

1) A **molecular ion**, $M^+_{(g)}$, is formed when the bombarding electrons remove 1 electron from the molecule. This gives the peak in the spectrum with the **highest mass** (furthest to the right, ignoring isotopes). The mass of M^+ gives $\mathbf{M_r}$ for the molecule, e.g. $CH_3CH_2OH^+$ has $M_r = 46$.

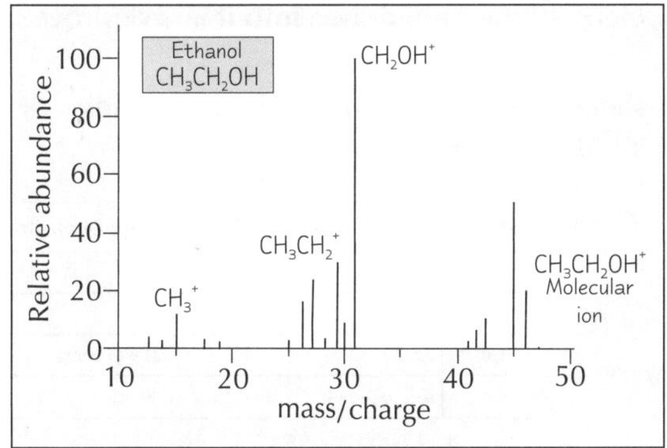

2) But it's not that simple — bombarding with electrons makes some molecules break up into fragments. These all show up on the mass spectrum, making a **fragmentation pattern**. For ethanol, the fragments you get include: CH_3^+ ($M_r = 15$), $CH_3CH_2^+$ ($M_r = 29$) and CH_2OH^+ ($M_r = 31$). Fragmentation patterns are actually pretty cool because you can use them to identify **molecules** and even their **structure**. There's more on this in Unit 2.

...and it has many Uses

1) **Carbon dating** is a method for working out the age of carbon-based things. About 1 part in a billion of the carbon in living things is **carbon-14**. Carbon-14 is radioactive so it decays, but living things are constantly taking in fresh carbon, so the amount stays constant.
   As soon as something dies and stops taking up new CO_2, the proportion of carbon-14 in the thing starts going down. Scientists can use **mass spectrometry** to measure how much carbon-14 an ancient plank of wood, say, contains — and so how long since the tree was chopped down.
2) The **pharmaceutical industry** uses mass spectrometry to **identify** the compounds in possible new drugs, and to tell how long drugs stay in the body by finding molecules of the drug in blood and urine samples.
3) Athletes' urine is **tested for drugs** with mass spectrometry. It shows up the presence of **banned substances** such as anabolic steroids by their distinctive spectra.
4) **Probes to Mars** have carried small mass spectrometers to study the composition of the surface of Mars and to look for molecules that might suggest that life existed on the planet.

Practice Questions

Q1 A sample of argon is injected into a mass spectrometer. Outline the main things that happen.

Q2 On a mass spectrum of an element, what does the height of each peak represent?

Q3 Explain how a mass spectrum shows the relative molecular mass of a compound.

Q4 Describe two uses of mass spectrometry.

Exam Questions

Q1 Copper, Cu, exists in two main isotopic forms, ^{63}Cu and ^{65}Cu.

a) Calculate the relative atomic mass of Cu using the information from the mass spectrum. [2 marks]

b) Explain why the relative atomic mass of copper is not a whole number. [2 marks]

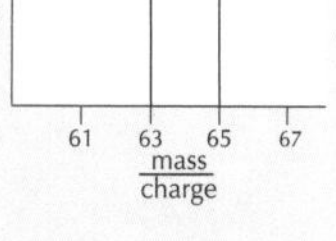

Q2 The percentage make-up of naturally occurring potassium is 93.11 % ^{39}K, 0.12 % ^{40}K and 6.77 % ^{41}K.

a) What method is used to determine the mass and abundance of each isotope? [1 mark]

b) Use the information to determine the relative atomic mass of potassium. [2 marks]

Relative abundance — I've got 6 aunts, 6 uncles, 17 nieces, 22 nephews...

Mass spectrometry is quite hard to get your head round, but stick at it — it could be worth loads of marks. And it's also one of those nice bits of 'real life' Chemistry that is actually used in real life. For example, it was used to work out the structure of the 'designer steroid', THG (tetrahydrogestrinone), when it first turned up in samples provided by athletes.

Electronic Structure

Those little electrons prancing about like bunnies are what chemistry's all about — they decide what'll react with what.

Electron Shells are Made Up of Sub-Shells and Orbitals

1) In the currently accepted models of the atom, electrons have **fixed energies**. They move around the nucleus in certain regions of the atom called **shells** or **energy levels**.
2) Each shell is given a number called the **principal quantum number**. The **further** a shell is from the nucleus, the **higher** its energy and the **larger** its principal quantum number.
3) This model helps to explain why electrons are **attracted** to the nucleus, but are not **drawn into it** and destroyed.
4) **Experiments** show that not all the electrons in a shell have exactly the same energy. The **atomic model** explains this — shells are divided up into **sub-shells** that have slightly different energies. The sub-shells have different numbers of **orbitals** which can each hold up to **2 electrons**.

This table shows the number of electrons that fit in each type of sub-shell.

Sub-shell	Number of orbitals	Maximum electrons
s	1	1 × 2 = 2
p	3	3 × 2 = 6
d	5	5 × 2 = 10
f	7	7 × 2 = 14

And this one shows the sub-shells and electrons in the first four energy levels.

Shell	Sub-shells	Total number of electrons	
1st	1*s*	2	= 2
2nd	2*s* 2*p*	2 + (3 × 2)	= 8
3rd	3*s* 3*p* 3*d*	2 + (3 × 2) + (5 × 2)	= 18
4th	4*s* 4*p* 4*d* 4*f*	2 + (3 × 2) + (5 × 2) + (7 × 2)	= 32

Orbitals Have Characteristic Shapes

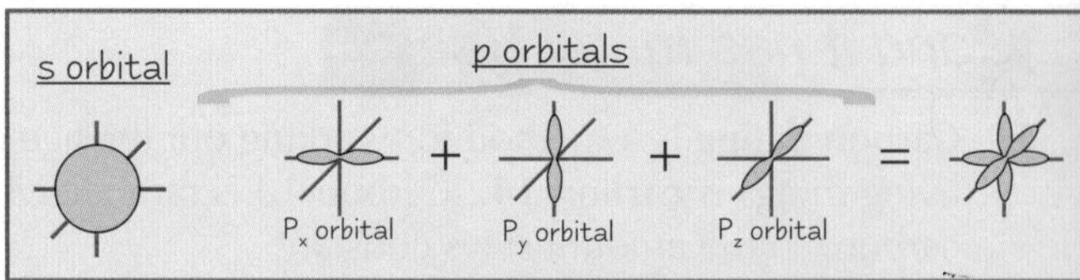

1) An orbital is the **bit of space** that an electron moves in. Orbitals within the same sub-shell have the **same energy**.
2) The orbitals are defined by **mathematical equations**. These equations are **models** for the ways electrons move.
3) s orbitals are **spherical**. p orbitals are **dumbbell-shaped**. There are 3 p orbitals at right angles to one another.
4) There are 5 **d orbitals**, so a d sub-shell can hold **10 electrons** — you don't need to know the shapes of d orbitals.
5) The diagrams of the shapes of orbitals are simplified versions of **graphs** of the equations describing the orbitals. These graphs are called **electron density plots**.

Work Out Electron Configurations by Filling the Lowest Energy Levels First

You can figure out most electronic configurations pretty easily, so long as you know a few simple rules —

1) Electrons fill up the **lowest** energy sub-shells first.

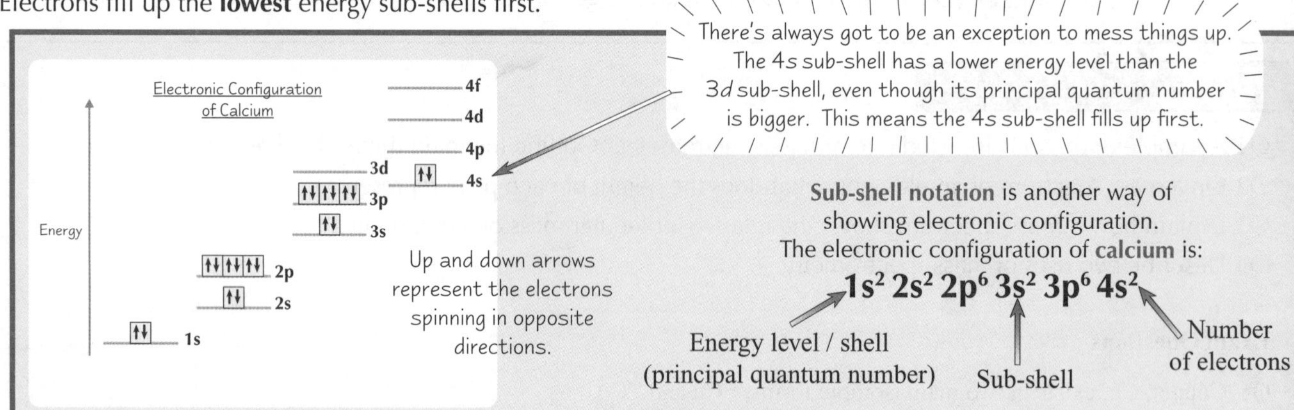

There's always got to be an exception to mess things up. The 4*s* sub-shell has a lower energy level than the 3*d* sub-shell, even though its principal quantum number is bigger. This means the 4*s* sub-shell fills up first.

Up and down arrows represent the electrons spinning in opposite directions.

Sub-shell notation is another way of showing electronic configuration. The electronic configuration of **calcium** is:

$$1s^2\ 2s^2\ 2p^6\ 3s^2\ 3p^6\ 4s^2$$

2) Electrons fill orbitals **singly** before they start sharing.

	1s	2s	2p
Nitrogen	↑↓	↑↓	↑ ↑ ↑
Oxygen	↑↓	↑↓	↑↓ ↑ ↑

See page 32 for more on the s and p block.

3) For the configuration of **ions** from the **s** and **p** blocks of the periodic table, just **remove or add** the electrons to or from the highest energy occupied sub-shell. E.g. $Mg^{2+} = 1s^2\ 2s^2\ 2p^6$, $Cl^- = 1s^2\ 2s^2\ 2p^6\ 3s^2\ 3p^6$

Watch out — **inert gas symbols**, like that of argon (Ar), are sometimes used in configurations. For example, calcium ($1s^2\ 2s^2\ 2p^6\ 3s^2\ 3p^6\ 4s^2$) can be written as $[Ar]4s^2$, where $[Ar] = 1s^2\ 2s^2\ 2p^6\ 3s^2\ 3p^6$.

Electronic Structure

Chromium and Copper Behave Unusually

Chromium (Cr) and **copper** (Cu) are badly behaved. They donate one of their **4s** electrons to the **3d sub-shell**. It's because they're happier with a **more stable** full or half-full d sub-shell.

Cr atom (24 e^-): $1s^2\ 2s^2\ 2p^6\ 3s^2\ 3p^6\ 3d^5\ 4s^1$ Cu atom (29 e^-): $1s^2\ 2s^2\ 2p^6\ 3s^2\ 3p^6\ 3d^{10}\ 4s^1$

Electronic Structure Decides the **Chemical Properties** of an Element

The number of **outer shell electrons** decides the chemical properties of an element.

1) The **s block** elements (Groups 1 and 2) have 1 or 2 outer shell electrons. These are easily **lost** to form positive ions with an **inert gas configuration**.
   E.g. Na — $1s^2\ 2s^2\ 2p^6\ 3s^1$ → Na^+ — $1s^2\ 2s^2\ 2p^6$
   (the electron configuration of neon).
2) The elements in Groups 5, 6 and 7 (in the p block) can **gain** 1, 2 or 3 electrons to form negative ions with an **inert gas configuration**.
   E.g. O — $1s^2\ 2s^2\ 2p^4$ → O^{2-} — $1s^2\ 2s^2\ 2p^6$.
   Groups 4 to 7 can also **share** electrons when they form covalent bonds.
3) Group 0 (the inert gases) have **completely filled** s and p sub-shells and don't need to bother gaining, losing or sharing electrons — their full sub-shells make them **inert**.
4) The **d block elements** (transition metals) tend to **lose** s and d electrons to form positive ions.

Sub-shells and the periodic table

1s 1s
2s 3s 4s 5s 6s 7s
3d 4d 5d
2p 3p 4p 5p 6p

Practice Questions

Q1 Write down the sub-shells in order of increasing energy up to 4*f*.

Q2 How many electrons would full s, p and d sub-shells contain?

Q3 Draw diagrams to show the shapes of an s and a p orbital.

Q4 What is strange about the way that chromium fills up its orbitals?

Exam Questions

Q1 Potassium can react with oxygen to form potassium oxide, K_2O.

a) Using sub-shell notation, give the electron configuration of the K atom and K^+ ion. [2 marks]

b) Using arrow-in-box notation, give the electron configuration of the oxygen atom. [2 marks]

c) Explain why it is mainly the outer shell electrons, not those in the inner shells, which determine the chemistry of potassium and oxygen. [2 marks]

Q2 This question concerns electron configurations in atoms and ions.

a) What is the electron configuration of manganese, Mn? [1 mark]

b) Identify the element with the 4th shell configuration of $4s^2\ 4p^2$. [1 mark]

c) Suggest the identity of an atom, a positive ion and a negative ion with the electron configuration $1s^2\ 2s^2\ 2p^6\ 3s^2\ 3p^6$. [3 marks]

d) Using electron-in-boxes notation, give the electron configuration of the Al^{3+} ion. [2 marks]

She shells sub-sells on the shesore...

The way electrons fill up the orbitals is kind of like how strangers fill up seats on a bus. Everyone tends to sit in their own seat till they're forced to share. Except for the huge, scary, smelly man who comes and sits next to you. Make sure you learn the order the sub-shells are filled up, so you can write electron configurations for any atom they throw at you.

Ionisation Energies

This page gets a trifle brain-boggling, so I hope you've got a few aspirin handy...

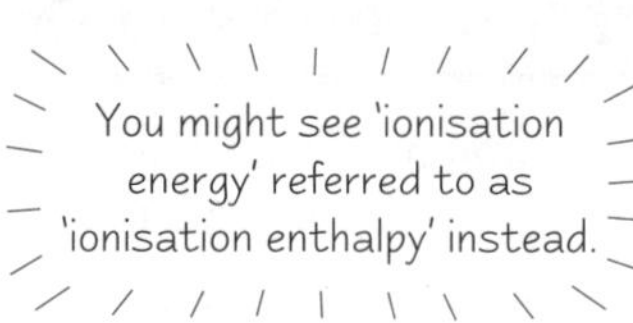

Ionisation is the Removal of One or More Electrons

When electrons have been removed from an atom or molecule, it's been **ionised**.
The energy you need to remove the first electron is called the **first ionisation energy**:

> The **first ionisation energy** is the energy needed to remove 1 electron from **each atom** in **1 mole** of **gaseous** atoms to form 1 mole of gaseous 1+ ions.

You have to put energy **in** to ionise an atom or molecule, so it's an **endothermic process** — there's more about endothermic processes on page 18.

You can write **equations** for this process — here's the equation for the **first ionisation of oxygen**:

$$O_{(g)} \rightarrow O^{+}_{(g)} + e^{-} \qquad \textbf{1st ionisation energy = +1314 kJ mol}^{-1}$$

Here are a few rather important points about ionisation energies:

1) You **must** use the gas state symbol, **(g)**, because ionisation energies are measured for gaseous atoms.
2) Always refer to **1 mole** of atoms, as stated in the definition, rather than to a single atom.
3) The **lower** the ionisation energy, the **easier** it is to form an ion.

The Factors Affecting Ionisation Energy are...

Nuclear Charge

The **more protons** there are in the nucleus, the more positively charged the nucleus is and the **stronger the attraction** for the electrons.

Attraction falls off very **rapidly with distance**. An electron **close** to the nucleus will be **much more** strongly attracted than one further away.

As the number of electrons **between** the outer electrons and the nucleus **increases**, the outer electrons feel less attraction towards the nuclear charge. This lessening of the pull of the nucleus by inner shells of electrons is called **shielding (or screening)**.

> A **high ionisation energy** means there's a **high attraction** between the **electron** and the **nucleus**.

There are Trends in First Ionisation Energies

1) The first ionisation energies of elements **down a group** of the periodic table **decrease**. Check out page 76 for why.
2) The first ionisation energies of elements **across a period generally increase**. But I do say **generally** — there's a bit more to it than that. Page 33 explains it in wondrous detail.

Successive Ionisation Energies Involve Removing Additional Electrons

1) You can remove **all** the electrons from an atom, leaving only the nucleus. Each time you remove an electron, there's a **successive ionisation energy**.
2) The definition for the **second ionisation energy** is —

> The **second ionisation energy** is the energy needed to remove 1 electron from **each ion** in **1 mole** of **gaseous** 1+ ions to form 1 mole of gaseous 2+ ions.

And here's the equation for the **second ionisation of oxygen** :

$$O^{+}_{(g)} \rightarrow O^{2+}_{(g)} + e^{-} \qquad \textbf{2nd ionisation energy = +3388 kJ mol}^{-1}$$

Ionisation Energies

Successive Ionisation Energies Show **Shell Structure**

A **graph** of successive ionisation energies (like this one for sodium) provides evidence for the **shell structure** of atoms.

1) **Within each shell**, successive ionisation energies **increase**. This is because electrons are being removed from an **increasingly positive ion** — there's **less repulsion** amongst the remaining electrons, so they're **held more strongly** by the nucleus.
2) The **big jumps** in ionisation energy happen when a new shell is broken into — an electron is being removed from a shell **closer** to the nucleus.

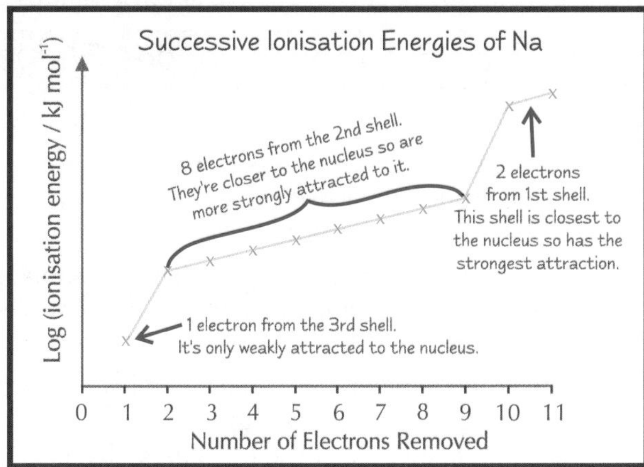

1) Graphs like this can tell you which **group** of the periodic table an element belongs to. Just count **how many electrons are removed** before the first big jump to find the group number.

   E.g. In the graph for sodium, **one electron** is removed before the first big jump — sodium is in **group 1**.

2) These graphs can be used to predict the **electronic structure** of an element. Working from **right to left**, count how many points there are before each big jump to find how many electrons are in each shell, starting with the first.

   E.g. The graph for sodium has **2 points** on the right-hand side, then a jump, then **8 points**, a jump, and **1 final point**. Sodium has **2 electrons** in the first shell, **8** in the second and **1** in the third.

Practice Questions

Q1 Define first ionisation energy and give an equation as an example.
Q2 Describe the three main factors that affect ionisation energies.
Q3 When an atom is ionised, does it release or absorb energy?
Q4 How is ionisation energy related to the force of attraction between an electron and the nucleus of an atom?

Exam Questions

Q1 This table shows the nuclear charge and first ionisation energy for four elements.

Element	B	C	N	O
Charge of Nucleus	+5	+6	+7	+8
1st Ionisation Energy ($kJ\ mol^{-1}$)	801	1087	1402	1314

a) Write an equation, including state symbols, to represent the first ionisation energy of carbon (C). [2 marks]
b) In these four elements, what is the relationship between nuclear charge and first ionisation energy? [1 mark]
c) Explain why nuclear charge has this effect on first ionisation energy. [2 marks]

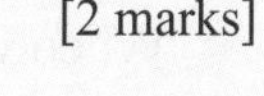

Q2 This graph shows the successive ionisation energies of a certain element.
a) To which group of the periodic table does this element belong? [1 mark]
b) Give two reasons why it takes more energy to remove each successive electron. [2 marks]
c) What causes the sudden increases in ionisation energy? [1 mark]
d) What is the total number of shells of electrons in this element? [1 mark]

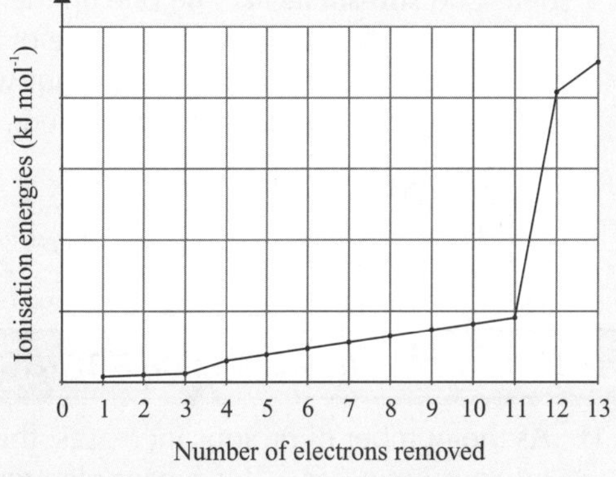

Shirt crumpled — ionise it...

When you're talking about ionisation energies in exams, always use the three main factors — shielding, nuclear charge and distance from nucleus. Recite the definition of the first and second ionisation energies to yourself until the men in white coats get to you. Then stop. I bet you can't wait until you get to that other stuff on ionisation energy I told you about.

Periodic Properties

A 'periodic property' is a physical or chemical property of elements that changes gradually as you move across a period, and the gradual change is repeated in each period. E.g. metal to non-metal is a trend that occurs going from left to right. You might have to explain why various properties change in the way they do, so read and enjoy...

The **Periodic Table** arranges Elements by **Proton Number**

1) The periodic table is arranged into **periods** (rows) and **groups** (columns), by atomic (proton) number.
2) All the elements **within a period** have the same number of **electron shells** (if you don't worry about s and p sub-shells) E.g. the elements in Period 2 have 2 electron shells.
3) All the elements **within a group** have the **same number** of electrons in their **outer shell** — so they have **similar properties**.
4) The **group number** tells you the number of electrons in the outer shell, e.g. Group 1 elements have 1 electron in their outer shell, Group 4 elements have 4 electrons and so on...

	Group 1	Group 2											Group 3	Group 4	Group 5	Group 6	Group 7	Group 0
								1 H Hydrogen 1										4 He 2
2	7 Li 3	9 Be 4											11 B 5	12 C 6	14 N 7	16 O 8	19 F 9	20 Ne 10
3	23 Na 11	24 Mg 12											27 Al 13	28 Si 14	31 P 15	32 S 16	35.5 Cl 17	40 Ar 18
4	39 K 19	40 Ca 20	45 Sc 21	48 Ti 22	51 V 23	52 Cr 24	55 Mn	56 Fe 26	59 Co 27	59 Ni 28	64 Cu 29	65 Zn 30	70 Ga 31	73 Ge 32	75 As 33	79 Se 34	80 Br 35	84 Kr 36
5	86 Rb 37	88 Sr 38	89 Y 39	91 Zr 40	93 Nb 41	96 Mo 42	99 Tc 43	101 Ru 44	103 Rh 45	106 Pd 46	108 Ag 47	112 Cd 48	115 In 49	119 Sn 50	122 Sb 51	128 Te 52	127 I 53	131 Xe 54
6	133 Cs 55	137 Ba 56	57-71 Lanthanides	179 Hf 72	181 Ta 73	184 W 74	186 Re 75	190 Os 76	192 Ir 77	195 Pt 78	197 Au 79	201 Hg 80	204 Tl 81	207 Pb 82	209 Bi 83	210 Po 84	210 At 85	222 Rn 86
7	223 Fr 87	226 Ra 88	89-103 Actinides															

You can use the Periodic Table to work out **Electron Configurations**

The periodic table can be split into an **s block**, **d block** and **p block** like this: Doing this shows you which sub-shells all the electrons go into.

See page 28 if this sub-shell malarkey doesn't ring a bell.

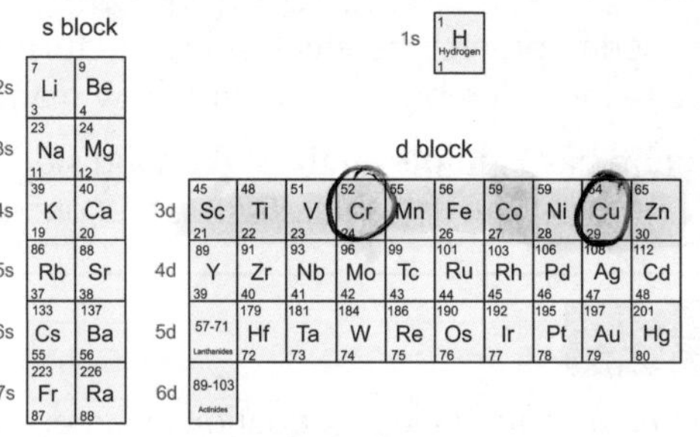

	p block					4 He 2
2p	11 B 5	12 C 6	14 N 7	16 O 8	19 F 9	20 Ne 10
3p	27 Al 13	28 Si 14	31 P 15	32 S 16	35.5 Cl 17	40 Ar 18
4p	70 Ga 31	73 Ge 32	75 As 33	79 Se 34	80 Br 35	84 Kr 36
5p	115 In 49	119 Sn 50	122 Sb 51	128 Te 52	127 I 53	131 Xe 54
6p	204 Tl 81	207 Pb 82	209 Bi 83	210 Po 84	210 At 85	222 Rn 86

1) The **s-block** elements have an outer shell electron configuration of s^1 or s^2.

Examples: Lithium ($1s^2$ $\mathbf{2s^1}$) and magnesium ($1s^2\ 2s^2\ 2p^6$ $\mathbf{3s^2}$)

2) The **p-block** elements have an outer shell configuration of s^2p^1 to s^2p^6.

Example: Chlorine ($1s^2\ 2s^2\ 2p^6$ $\mathbf{3s^2\ 3p^5}$)

3) The **d-block** elements have electron configurations in which d sub-shells are being filled.

Example: Cobalt ($1s^2\ 2s^2\ 2p^6\ 3s^2\ 3p^6$ $\mathbf{3d^7}$ $4s^2$)

Even though the 3d sub-shell fills last in cobalt, it's not written at the end of the line.

When you've got the periodic table **labelled** with the **shells** and **sub-shells** like the one up there, it's pretty easy to read off the electron structure of any element by starting at the top and working your way across and down until you get to your element.

A wee apology...
This bit's really hard to explain clearly in words. If you're confused, just look at the examples until you get it...

Example: Electron structure of phosphorus (P):

Period 1 — $1s^2$ ← Complete sub-shells
Period 2 — $2s^2\ 2p^6$ ← Complete sub-shells
Period 3 — $3s^2\ 3p^3$ ← Incomplete outer sub-shell

So the full electron structure of phosphorus is: $1s^2\ 2s^2\ 2p^6\ 3s^2\ 3p^3$

Atomic Radius **Decreases** across a Period

1) As the number of protons increases, the **positive charge** of the nucleus increases. This means electrons are **pulled closer** to the nucleus, making the atomic radius smaller.
2) The extra electrons that the elements gain across a period are added to the **outer energy level** so they don't really provide any extra shielding effect (shielding works with inner shells mainly).

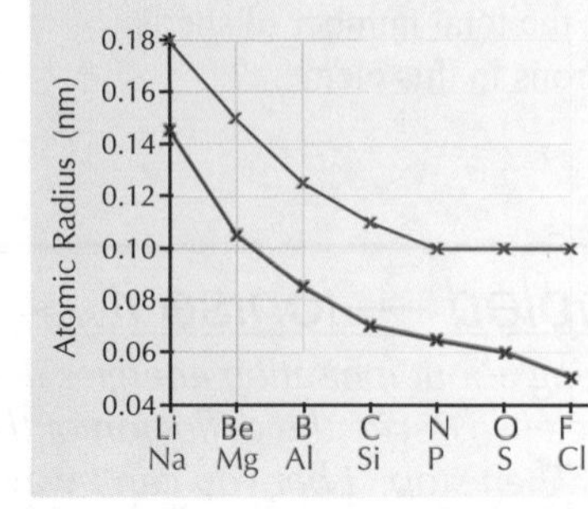

Periodic Properties

Ionisation Energy **Increases** across a Period

Don't forget — there are **3 main things** that affect the size of ionisation energies:

1) Atomic radius — the further the outer shell electrons are from the positive nucleus, the less they'll be attracted towards the nucleus. So, the ionisation energy will be **lower**.
2) Nuclear charge — the **more protons** there are in the nucleus, the more it'll attract the outer electrons — it'll be harder to remove the electrons, so the ionisation energy will be **higher**.
3) Electron shielding — the inner electron shells **shield** the outer shell electrons from the attractive force of the nucleus. Because more inner shells means more shielding, the ionisation energy will be **lower**.

The graph below shows the first ionisation energies of the elements in **Periods 2 and 3**.

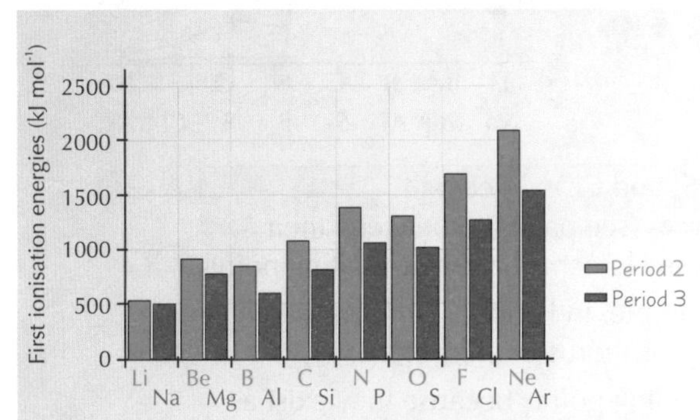

1) As you **move across** a period, the general trend is for the ionisation energies to **increase** — i.e. it gets harder to remove the outer electrons.
2) This is because the number of protons is increasing, which means a stronger **nuclear attraction**.
3) All the extra electrons are at **roughly the same** energy level, even if the outer electrons are in different orbital types.
4) This means there's generally little **extra shielding** effect or **extra distance** to lessen the attraction from the nucleus.
5) But, there are **small drops** between Groups 2 and 3, and 5 and 6. Tell me more, I hear you cry. Well, all right then...

Electronic Structure Explains the Drop between Groups 2 and 3

Be	$1s^2 2s^2$	1st ionisation energy = 900 kJ mol^{-1}
B	$1s^2 2s^2 2p^1$	1st ionisation energy = 799 kJ mol^{-1}

1) The crucial difference here is that while beryllium's outer electron is in the **2s sub-shell**, boron's outer electron is in the **2p sub-shell**.
2) The 2p sub-shell has a **slightly higher** energy than the 2s. So boron's outer electron is, on average, just a little bit **further** from the nucleus.
3) That little extra distance means there's not as much attraction between the nucleus and the outer electron. And the **$2s^2$ electrons** give the 2p sub-shell some **partial shielding**.
4) The combination of these two factors **overrides** the effect of the increased nuclear charge, resulting in a slight **drop** in ionisation energy.

The same thing happens with magnesium and aluminium, but in the **third** energy level: Aluminium has one **3p electron**, which is, on average, slightly further away from the nucleus than the 3s electrons, and also **partially shielded** by the 3s electrons. These two factors override the larger nuclear charge, so the first ionisation energy of aluminium is **lower** than magnesium's.

The Drop between Groups 5 and 6 is due to **Electron Repulsion**

N	$1s^2 2s^2 2p^3$	1st ionisation energy = 1400 kJ mol^{-1}
O	$1s^2 2s^2 2p^4$	1st ionisation energy = 1310 kJ mol^{-1}

1) This time the outer electrons for both elements are in the **same sub-shell**, so there's no difference in the shielding or the distance from the nucleus. But the key is how the sub-shells are filled.
2) The 2p sub-shell has three **orbitals** (see p28), and nitrogen has one electron in each of them. One of oxygen's 2p orbitals has **two electrons** in it. These two electrons **repel** each other — making it **easier** to remove one of them.
3) The first ionisation energy of **sulfur** is lower than that of **phosphorus** for exactly the same reason (but in the third energy level, of course).

Periodic Properties

Melting Points Are Linked to **Bond Strength** and **Structure**

Periods 2 and 3 show similar trends in their melting points.
These trends are linked to changes in **structure** and **bond strength**.

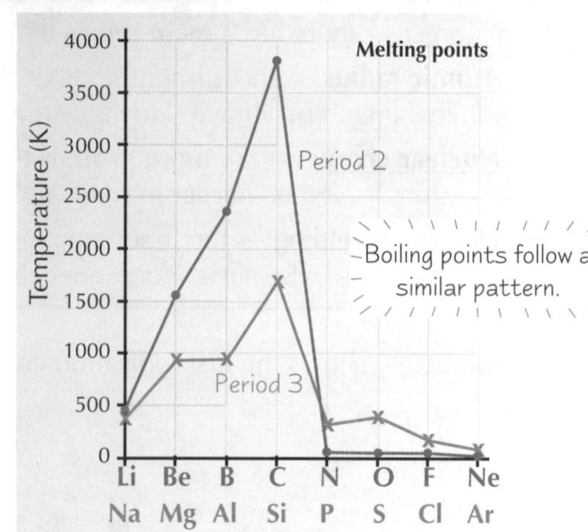

1) For the **metals** (Li and Be, Na, Mg and Al), melting points **increase** across the period because the **metal-metal bonds** get stronger. The bonds get stronger because the metal ions have an increasing number of **delocalised electrons** and a decreasing **radius**. This leads to a higher **charge density**, which attracts the ions together more strongly. See p43 for charge density.
2) The elements with **macromolecular** structures (p46) have **strong covalent bonds** linking all their atoms together. **A lot** of energy is needed to break these bonds. So, for example, carbon (as graphite or diamond) and silicon have the **highest** melting points in their periods. (The carbon data in the graph opposite is for graphite — diamond has an even higher melting point. But neither of them actually melts or boils at atmospheric pressure, they sublime from solid to gas.)
3) Next come the **simple molecular substances** (N_2, O_2 and F_2, P_4, S_8 and Cl_2) — see p44. Their melting points depend upon the strength of the **London forces** (see page 70) between their molecules. London forces are weak and easily overcome, so these elements have **low** melting points.
4) More atoms in a molecule mean stronger London forces. For example, in Period 3 sulfur is the **biggest molecule** (S_8), so it's got higher melting and boiling points than phosphorus or chlorine.
5) The noble gases (neon and argon) have the **lowest** melting and boiling points because they exist as **individual atoms** (they're monatomic) resulting in **very weak** London forces.

Practice Questions

Q1 Which elements in Period 3 are found in the s block of the periodic table?

Q2 Explain the meaning of the term 'periodic property'.

Q3 What happens to the ionisation energy as you move across a period?

Q4 Which factors control the trend for ionisation energy in a period?

Q5 Which element in Period 3 has the highest melting point? Which has the highest boiling point?

Exam Questions

Q1 Explain why the melting point of magnesium is higher than that of sodium. [3 marks]

Q2 The first ionisation energies of the elements lithium to neon are given here in kJ mol^{-1}:

Li	Be	B	C	N	O	F	Ne
519	900	799	1090	1400	1310	1680	2080

a) Explain why the ionisation energies show an overall tendency to increase across the period. [3 marks]

b) Explain the irregularities in this trend for:
(i) Boron (ii) Oxygen [4 marks]

Q3 Explain why the first ionisation energy of neon is greater than that of sodium. [2 marks]

Q4 This table shows the melting points of the Period 3 elements.

Element	Na	Mg	Al	Si	P	S	Cl	Ar
Melting point / K	371	923	933	1680	317	392	172	84

In terms of structure and bonding explain why:
a) silicon has a high melting point. [2 marks]
b) sulfur has a higher melting point than phosphorus. [2 marks]

Periodic trends — my mate Dom's always a decade behind...

*He still thinks Oasis, Blur and REM are the best bands around. The sad muppet. But not me. Oh no sirree, I'm up with the times — April Lavigne... Linkin' Pork... Christina Agorrilla. I'm hip, I'm with it. Da ga da ga da ga da ga.. ooaarrr ooup **

** Obscure reference to Austin Powers: International Man of Mystery. You should watch it — it's better than doing Chemistry.*

Ionic Bonding

Every atom's aim in life is to have a full outer shell of electrons. Once they've managed this, that's it — they're happy.

Compounds are Atoms of Different Elements Bonded Together

1) When different elements join or bond together, you get a **compound**.
2) There are two main types of bonding in compounds — **ionic** and **covalent**. You need to make sure you've got them **both** totally sussed.

E.g. when the elements hydrogen (H_2) and oxygen (O_2) combine, the compound water (H_2O) is formed.

Ionic Bonding is when Ions are Stuck Together by Electrostatic Attraction

1) Ions are formed when electrons are **transferred** from one atom to another.
2) The simplest ions are single atoms which have either lost or gained 1, 2 or 3 electrons so that they've got a **full outer shell**. Here are some examples:

A sodium atom (Na) **loses** 1 electron to form a sodium ion (Na^+)	$Na \rightarrow Na^+ + e^-$
A magnesium atom (Mg) **loses** 2 electrons to form a magnesium ion (Mg^{2+})	$Mg \rightarrow Mg^{2+} + 2e^-$
A chlorine atom (Cl) **gains** 1 electron to form a chloride ion (Cl^-)	$Cl + e^- \rightarrow Cl^-$
An oxygen atom (O) **gains** 2 electrons to form an oxide ion (O^{2-})	$O + 2e^- \rightarrow O^{2-}$

3) **Positive** ions, like Na^+, are called **cations**. That's because they're attracted to cathodes (negative electrodes). **Negative** ions, like Cl–, are called **anions** — yep, because they find themselves strangely drawn to anodes.
4) **Electrostatic attraction** holds cations and anions together — it's **very** strong. When atoms are held together like this, it's called **ionic bonding**.

Sodium Chloride and Magnesium Oxide are Ionic Compounds

1) The formula of sodium chloride is **NaCl**. It just tells you that sodium chloride is made up of **Na^+ ions** and **Cl^- ions** (in a 1:1 ratio).
2) You can use '**dot-and-cross**' diagrams to show how ionic bonding works in sodium chloride —

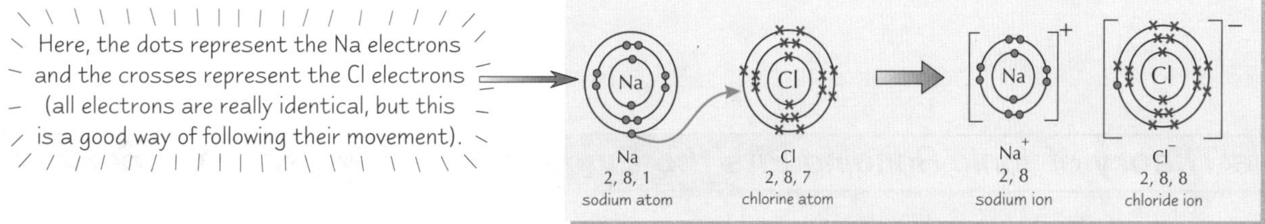

3) **Magnesium oxide**, MgO, is another good example:

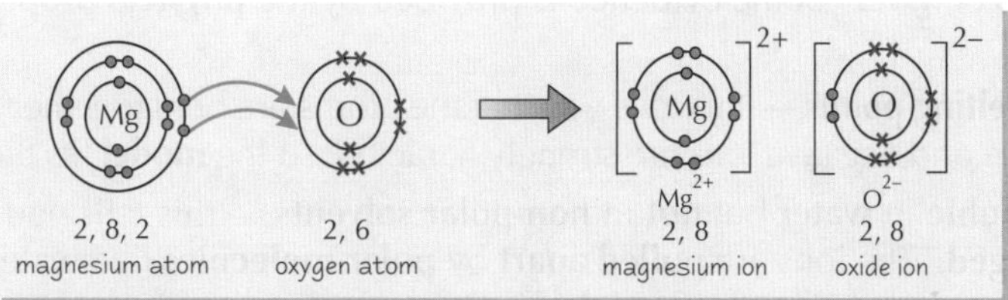

Dot (cross)

The positive charges in the compound **balance** the negative charges exactly — so the total overall charge is **zero**. This is a dead handy way of checking the formula.

- In **NaCl**, the single positive charge on the Na^+ ion balances the single negative charge on the Cl^- ion.
- In magnesium chloride, **$MgCl_2$**, the 2+ charge on the Mg^{2+} ion balances the two individual negative charges on the two Cl^- ions.

Ionic Bonding

Ionic Crystals are Giant Lattices of Ions

1) Ionic crystals (e.g. crystals of common salt) are giant lattices of ions. A **lattice** is just a **regular structure**.
2) The structure's called '**giant**' because it's made up of the same basic unit repeated over and over again — an ionic crystal doesn't have a set number of atoms in it.
3) In **sodium chloride**, the Na^+ and Cl^- ions are packed together in a **cubic** structure.

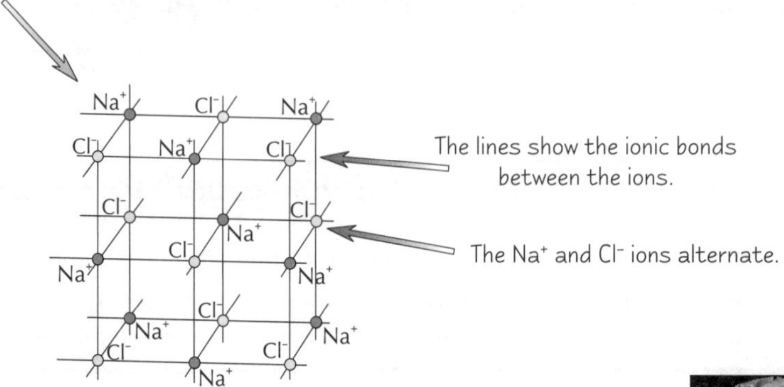

4) Different ionic compounds have differently shaped structures. There are quite a few different arrangements — but they're all **giant lattices**.

Oh, please. How predictable.

Elements in the Same Group Form Ions with the Same Charge

You should know this already, but better safe than sorry...

1) You **don't** have to remember what ion each **element** forms — nope, you just look at the periodic table.
2) Elements in the same **group** all have the same number of **outer electrons**. So they have to **lose or gain** the same number to get the full outer shell that they're aiming for. And this means that they form ions with the **same charges**.

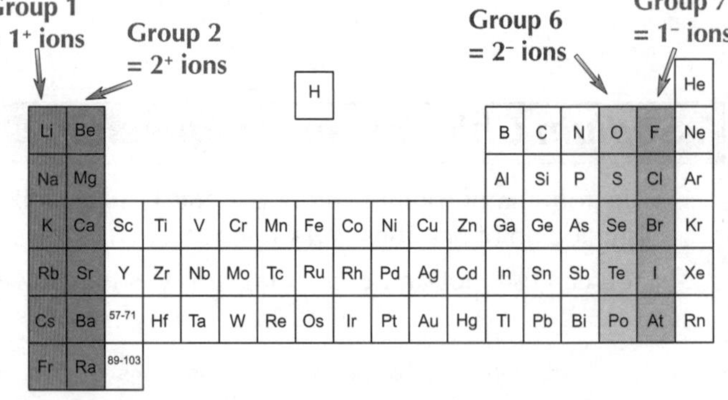

The Theory of Ionic Bonding Fits the Evidence from Physical Properties

So that was the **theory** behind what an ion is and how ionic bonding happens.

Scientists develop **models** of ionic bonding based on **experimental evidence** — they're an attempt to **explain observations** about how ionic compounds behave. Some evidence is provided by the **physical properties** of ionic compounds:

1) They have **high melting points** — this tells you that the atoms are held together by a **strong attraction**. Positive and negative ions are strongly attracted, so the **model** fits the **evidence**.
2) They are often **soluble** in **water** but **not** in **non-polar solvents** — this tells you that the particles are **charged**. The ions are **pulled apart** by **polar molecules** like water, but **not** by **non-polar** molecules. Again, the **model** of ionic structures fits this evidence.
3) Ionic compounds **don't conduct electricity** when they're **solid** — but they **do** when they're **molten or dissolved**. This supports the idea that there are ions, which are **fixed** in position by strong ionic bonds in a solid, but are **free to move** (and carry a charge) as a liquid or in a solution.

Ionic Bonding

Not All **Ions** are Made from **Single Atoms**

Ions such as Na^+, Cl^-, Mg^{2+}, O^{2-} and Fe^{3+} are formed from **single atoms.**

There are lots of ions that are made up of a group of atoms with an overall charge. These are called **compound ions**. You'll come across some of them pretty often in AS Chemistry, so you need to know their formulas.

Nitrate	Carbonate	Sulfate	Ammonium
NO_3^-	CO_3^{2-}	SO_4^{2-}	NH_4^+

Practice Questions

Q1 What's a compound?

Q2 Draw a 'dot and cross' diagram showing the bonding between magnesium and oxygen.

Q3 What type of force holds ionic substances together?

Q4 What is a giant lattice?

Q5 What sort of ions do Group 6 elements form?

Q6 Do ionic compounds conduct electricity? Explain your answer.

Q7 What's the formula for a sulfate ion?

Exam Questions

Q1 a) Draw a labelled diagram to show the structure of sodium chloride. [3 marks]

b) What is the name of this type of structure? [1 mark]

c) Would you expect sodium chloride to have a high or a low melting point?
Explain your answer. [4 marks]

Q2 a) Ions can be formed by electron transfer. Explain this and give an example of a positive and a negative ion. [3 marks]

b) Solid lead bromide does not conduct electricity, but molten lead bromide does.
Explain this with reference to ionic bonding. [5 marks]

Q3 a) Explain, in terms of electron transfer, what happens when sodium reacts with fluorine to form sodium fluoride. [3 marks]

b) Draw electron configuration diagrams of a sodium ion and a fluoride ion. [4 marks]

Q4 The table below summarises some properties of compound Q.

Melting point	Solubility in water	Conductivity of solution	Conductivity of solid	Conductivity of liquid
High	High	Good	Very poor	Good

Explain how the data supports the suggestion that compound Q is ionic. [4 marks]

Any old ion, any old ion, any any any old ion...

This stuff's easy marks. Just make sure you can draw dot and cross diagrams showing the bonding in ionic compounds and you're more or less sorted. Remember — atoms are lazy. It's easier to lose two electrons to get a full shell than it is to gain six, so that's what an atom's going to do. Practise drawing sodium chloride too, and don't stop till you're perfect.

More on Ions

Just when you thought it was safe to go back to revision — another page on ions.

Ions Are *Smaller* than Atoms For *Metals* But *Larger* For *Non-Metals*

Ions aren't the same size as the atom they're formed from...

1) The ionic radius of a **metallic** element is **smaller** than the **atomic radius.**
2) Metals **lose electrons** when they form **ions**, so the **positive charge** of the nucleus is **larger** than the **negative charge** in the electron cloud. This means that in an ion, the electrons are pulled closer.
3) Also, when positive ions are formed, the outer electron shell is usually emptied, meaning there are fewer shells. The outer shell is now **closer** to the nucleus and there's less **electron shielding**.
4) So, the outer electrons are attracted **more strongly** to the nucleus.

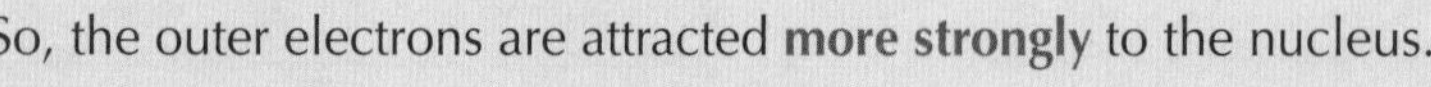

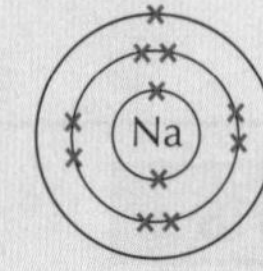

1) The ionic radius of a **non-metal** is **larger** than the **atomic radius.**
2) Non-metals **gain electrons** when they form **ions**. So there's a bigger negative charge in the **electron cloud** of the ion, which means there's **greater repulsion** between the electrons and the electron cloud **expands** a bit.

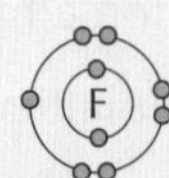

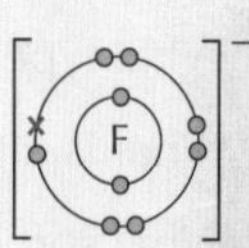

The *Size* of an *Ion* Depends On Its *Atomic Number* and *Charge*

There are two **trends** in ionic radii that you need to know about.

1) The **ionic radius increases** as you go **down a group**.

Ion	Ionic radius (nm)
Li^+	0.060
Na^+	0.095
K^+	0.133
Rb^+	0.148

All these **Group 1** ions have the **same charge**.

As you go down the group the **ionic radius increases** as the **atomic number increases.**

This is because **extra electron shells** are added.

Class 5B wished Mr Evans had come up with a simpler way to illustrate trends in ionic radius.

2) **Isoelectronic ions** are ions of different atoms with the **same number of electrons**. The **ionic radius** of a set of **isoelectronic ions decreases** as the **atomic number increases**.

Ion	Number of electrons	Number of protons	Ionic radius (nm)
N^{3-}	10	7	0.171
O^{2-}	10	8	0.140
F^-	10	9	0.136
Na^+	10	11	0.095
Mg^{2+}	10	12	0.065
Al^{3+}	10	13	0.050

As you go through this series of ions the number of **electrons stays the same**, but the number of **protons increases**.

This means that the electrons are **attracted** to the **nucleus** more strongly, pulling them in a little, so the **ionic radius decreases.**

More on Ions

There are Different Kinds of Evidence for the Existence of Ions

The **physical properties** of ionic compounds provide evidence for the existence of ions (see p36). There are few more bits of **evidence** you need to know about too.

1) The **migration of ions on wet filter paper** is evidence for the presence of charged particles.

When you **electrolyse** a **green** solution of **copper(II) chromate(VI)** the filter paper turns **blue** at the **cathode** and yellow at the anode.
Copper(II) ions are **blue** in solution and chromate(VI) ions are yellow.
Copper(II) chromate(VI) solution is **green** because it contains **both** ions.
When you pass a current through the solution, the **positive** ions move to the **cathode** and the negative ions move to the anode.

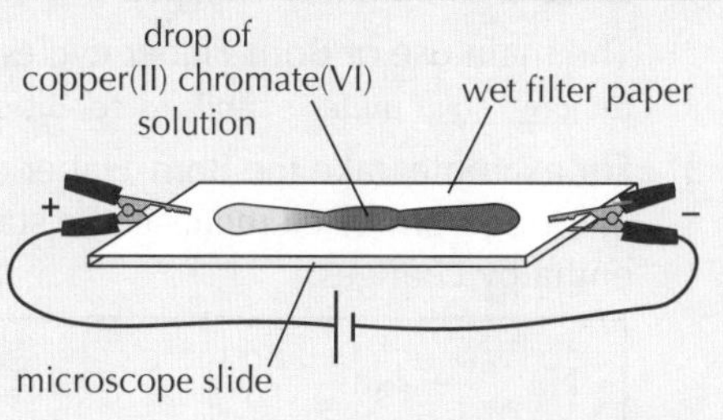

2) **Electron density maps** look like contour maps. The lines on the map join parts of the molecule that have the **same density** of **electrons**.
The electron density map of an **ionic** crystal shows that there are **spaces** between the ions where the density of electrons is **zero**. This shows that the atoms have **no shared electrons** — the bonding electrons have moved from one atom to the other.

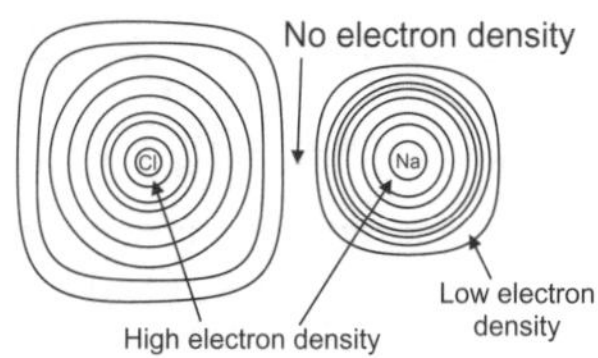

Electron density maps are made using X-ray crystallography. A beam of X-rays is fired at a crystal and the electrons in the crystal scatter the X-rays. The pattern of scattering gives a picture of the electron density throughout the crystal — this is the electron density map.

Practice Questions

Q1 What are isoelectronic ions?

Q2 Explain why an aluminium ion is smaller than a magnesium ion even though they are isoelectronic.

Q3 Explain how an electron density map of NaCl shows that electrons are not being shared between the atoms.

Exam Questions

Q1 a) The ions O^{2-} and Na^{+} have the same number of electrons as an element in Group 0. Which element? [1 mark]

b) Explain the differences in size between the two ions and the Group 0 atom. [4 marks]

Q2 The apparatus shown in the diagram is set up. After a while a purple streak is seen moving from the crystal to the anode.

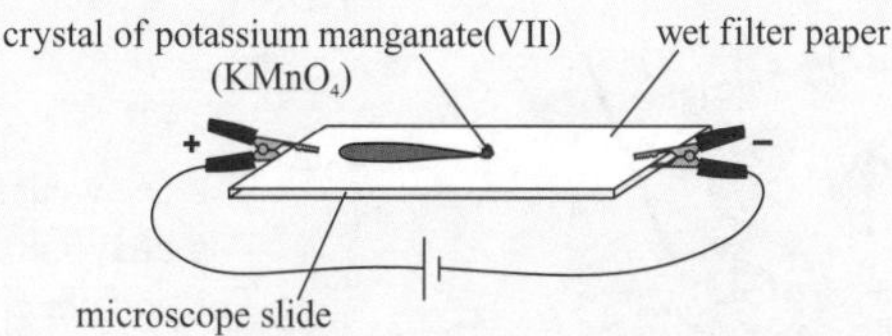

a) What does this experiment show about the colour of potassium ions? [1 mark]

b) How does this experiment show that ions are charged particles? [2 marks]

This sentence was printed using Copper(II) Chromate(VI) solution...

...which means that the left-hand side of this page must be positive, and the right-hand side must be negative. Weird. Anyhow, the migration of coloured ions is a great piece of evidence that you can use, say in an exam, to back up all this theory about what an ion is. And what is it that you need to back up a theory? Dedication? Nope. Evidence. That's what you need.

Ions and Born-Haber Cycles

Born-Haber cycles can seem a bit much at first, but stick with it and practise lots— the fog will clear. Promise.

Born-Haber Cycles Can Be Used to Calculate Lattice Energies

1) **Born-Haber cycles** show enthalpy changes when a **solid ionic compound** is formed from its **elements** in their standard states. They show two 'routes' — one direct and one indirect. From Hess's Law, both routes have the **same** total enthalpy change.
2) The main use of Born-Haber cycles is to calculate **lattice energies** (the energy change when gaseous ions form 1 mole of an ionic solid under standard conditions) — because lattice energies can't be found directly from experiments.
3) For example, take the Born-Haber cycle for the formation of **sodium chloride**. The **direct** way to form sodium chloride from its elements is the **standard enthalpy of formation**. The **indirect** route involves adding up all these enthalpy changes:

$Na_{(s)} \rightarrow Na_{(g)}$	**standard enthalpy of atomisation** of sodium metal, $\Delta H^{\ominus}_{at}$ $[Na_{(s)}]$	109 kJ mol^{-1}
$Na_{(g)} \rightarrow Na^{+}_{(g)} + e^{-}$	**first ionisation energy** of sodium, ΔH_{m1} $[Na_{(g)}]$ or E_{m1} $[Na_{(g)}]$	500 kJ mol^{-1}
$\frac{1}{2} Cl_{2\,(g)} \rightarrow Cl_{(g)}$	**standard enthalpy of atomisation** of chlorine gas, $\Delta H^{\ominus}_{at}$ $[Cl_{(g)}]$	121 kJ mol^{-1}
$Cl_{(g)} + e^{-} \rightarrow Cl^{-}_{(g)}$	**electron affinity** of chlorine, ΔH_{m1} $[Cl_{(g)}]$ or E_{aff} $[Cl_{(g)}]$ (The energy change when one mole of ions is formed from one mole of atoms.)	–370 kJ mol^{-1}
$Na^{+}_{(g)} + Cl^{-}_{(g)} \rightarrow NaCl_{(s)}$	**standard lattice energy** of sodium chloride, $\Delta H^{\ominus}_{latt}$ $[Na^{+}Cl^{-}_{(s)}]$	You're probably trying to find this.

State symbols are important in Born-Haber cycles — elements must be in their **standard states** at the beginning. (Remember, 'standard states' means their states at 298 K, 100 kPa pressure.)

The elements (in their standard states) that make up sodium chloride are sodium metal, $Na_{(s)}$, and chlorine gas $Cl_{2\,(g)}$.

So, here's the Born-Haber cycle for sodium chloride — you start reading it from the bottom:

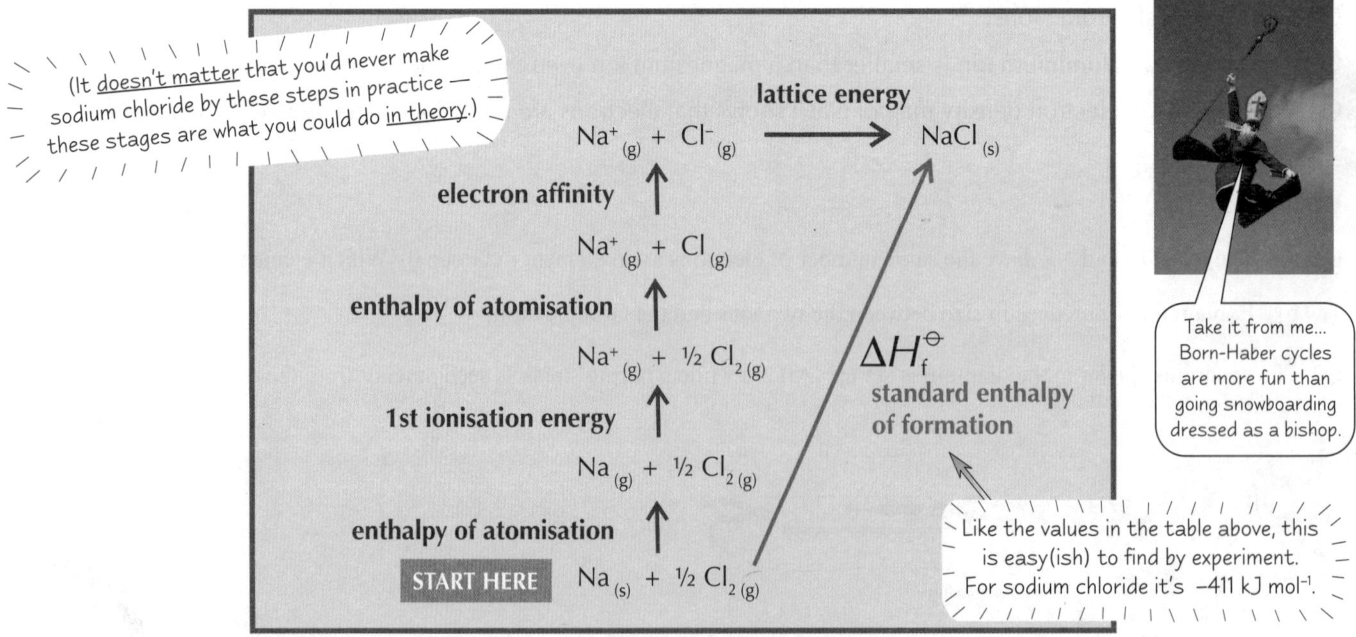

4) Using **Hess's Law**, the **direct route** (the **green arrow**) = the **indirect route** (the **purple** arrows). So to find the **lattice energy** from the Born-Haber cycle:

$$\Delta H^{\ominus}_{f} [NaCl_{(s)}] = \Delta H^{\ominus}_{at} [Na_{(s)}] + E_{m1} [Na_{(g)}] + \Delta H^{\ominus}_{at} [Cl_{(g)}] + E_{aff} [Cl_{(g)}] + \Delta H^{\ominus}_{latt} [Na^{+}Cl^{-}_{(s)}]$$

$-411 = 109 + 500 + 121 + (-370) + \Delta H^{\ominus}_{latt} [Na^{+}Cl^{-}_{(s)}]$

Now rearrange it — making sure you get the **signs** correct:

$\Delta H^{\ominus}_{latt} [Na^{+}Cl^{-}_{(s)}] = -411 - 109 - 500 - 121 - (-370)$ = **-771 kJ mol^{-1}**

Ions and Born-Haber Cycles

Born-Haber Cycles Can Show **Why Some Compounds Don't Exist**

1) If lots of energy is **released** during the formation of a compound, the compound you get is nice and **stable** — it doesn't 'want' to break up again. (It's a bit like when things roll down a hill — they stay at the bottom and don't go uphill again.)
2) Some compounds either **don't form** in the first place, or else they **break up** very quickly to form more stable elements and compounds. For example, **NaCl** is found all over the place, but you never get **$NaCl_2$**. Here's why:

> 1) The Born-Haber cycle for NaCl gives a **negative** enthalpy of formation — the formation of NaCl from its elements is an **exothermic** process — energy is released overall. So NaCl is **stable**.
> 2) However, $NaCl_2$ is a completely different matter. To make it, you need **Na^{2+}** ions....
> 3) So the Born-Haber cycle for $NaCl_2$ would include the **second ionisation energy** for sodium — a whopping **+4560 kJ mol^{-1}** (it needs so much energy because you're trying to take an electron from Na^+'s now full outer shell).
> 4) When you add everything up, the enthalpy of formation of **$NaCl_2$** is **positive**. So forming $NaCl_2$ is an **endothermic** process — you need to put loads of energy in. It just **doesn't happen** because it's **energetically unfavourable** — the compound wouldn't be stable.

3) Similarly, $MgCl_2$ is much more stable than either $MgCl_3$ or MgCl:

> - $MgCl_3$ can't exist because the huge third ionisation energy of Mg makes $\Delta H_f^\ominus$ **positive**.
> - MgCl and $MgCl_2$ both have negative enthalpies of formation, so both compounds **can** form. But $MgCl_2$ has $\Delta H_f^\ominus = -673$ kJ mol^{-1}, whereas MgCl has $\Delta H_f^\ominus = -111$ kJ mol^{-1}. This means that **more energy** is released by forming $MgCl_2$, so $MgCl_2$ is **more stable**. So if any MgCl forms in a chemical reaction, it immediately **disproportionates** (see p80) to form $MgCl_2$ and Mg. ⟹ $2MgCl \rightarrow MgCl_2 + Mg$

Practice Questions

Q1 What enthalpy change are Born Haber cycles usually used to calculate?

Q2 Give chemical equations for the following, including the state symbols of all the species present:

a) $\Delta H_{at}^\ominus$ [$K_{(s)}$] b) E_{m1} [$K_{(s)}$] c) E_{aff} [$I_{(g)}$] d) $\Delta H_{latt}^\ominus$ [$KI_{(s)}$] b) $\Delta H_f^\ominus$ [$MgO_{(s)}$]

Exam Questions

Q1 The enthalpy changes involved in the formation of calcium oxide are shown below.

Enthalpy of atomisation of calcium = +177 kJ mol^{-1}
First ionisation energy of calcium = +590 kJ mol^{-1}
Second ionisation energy of calcium = +1100 kJ mol^{-1}
Enthalpy of atomisation of oxygen = +249 kJ mol^{-1}
Electron affinity of an oxygen atom = −141 kJ mol^{-1}
Electron affinity of O^- = +790 kJ mol^{-1}
Lattice energy of calcium oxide = −3401 kJ mol^{-1}

a) Calculate the enthalpy of formation for calcium oxide using the information given above. [3 marks]

b) The electron affinity of the O^- ion is +790 kJ mol^{-1}. Explain why the electron affinity of O^- is positive. [2 marks]

Q2 Use the data below to calculate the lattice energy of magnesium chloride, $MgCl_2$.

Enthalpy of atomisation of magnesium = +148 kJ mol^{-1}
First ionisation energy of magnesium = +738 kJ mol^{-1}
Second ionisation energy of magnesium = +1451 kJ mol^{-1}
Enthalpy of atomisation of chlorine = +122 kJ mol^{-1}
Electron affinity of a chlorine atom = −349 kJ mol^{-1}
Enthalpy of formation of $MgCl_2$(s) = −641 kJ mol^{-1}

[4 marks]

I've always loved sheltering ships — people say I'm a Born-Haber...

There's a good reason why some compounds exist and not others, and it's all to do with $\Delta H_f^\ominus$. If that's a pretty meaningless squiggle as far as you're concerned, now's the time to make friends with it — and to understand all the other bits that make up a Born-Haber cycle. Oh, and don't forget to practise those nice calculations for lattice energy and whatnot.

Lattice Energies & Polarisation of Ions

And you thought you'd finished with lattice energies...

Theoretical Lattice Energies are Based on the **Ionic Model**

1) There are **two ways** to work out a lattice energy:
   - the **experimental** way — using **experimental enthalpy values** in a Born-Haber cycle (as on the previous page)
   - the **theoretical** way — doing some calculations based on the **purely ionic model** of a lattice
2) To work out a 'theoretical' lattice energy, you assume that all the ions are **spherical** and have their charge **evenly distributed** around them — a **purely ionic** lattice. Then you work out how strongly the ions are **attracted** to one another based on their charges, the distance between them and so on (you don't need to know the details of these calculations, fortunately — just what they're based on). That gives you a value for the energy change when the ions form the lattice.

Comparing Lattice Energies Can Tell You **'How Ionic'** an Ionic Lattice Is

For any one compound, the experimental and theoretical lattice energies are usually **different**.
How different they are tells you **how closely** the lattice **actually** resembles the 'purely ionic' model used for the theoretical calculations. For example, here are both lattice energy values for some **sodium halides**.

Compound	Lattice Energy ($kJ\ mol^{-1}$)	
	From experimental values in Born-Haber cycle	From theory
Sodium chloride	−771	−766
Sodium bromide	−742	−731
Sodium iodide	−698	−686

1) The experimental and theoretical values are a **pretty close match** — so you can say that these compounds fit the 'purely ionic' model (spherical ions with evenly distributed charge, etc.) very well.
2) This indicates that the structure of the lattice for these compounds is quite close to being **purely ionic**.

Right then, so far so good. Here are some more lattice energies, for **magnesium halides** this time:

Compound	Lattice Energy ($kJ\ mol^{-1}$)	
	From experimental values in Born-Haber cycle	From theory
Magnesium chloride	−2526	−2326
Magnesium bromide	−2440	−2097
Magnesium iodide	−2327	−1944

1) The **experimental** lattice energies are **bigger** than the theoretical values by a fair bit — 10% or so.
2) This tells you that the **bonding** is, in practice, **stronger** than the calculations from the ionic model predict.
3) The difference shows that the bonding in the magnesium halides **isn't** as close to 'purely ionic' as it is with sodium halides.
4) It tells you that the ionic bonds in the magnesium halides are more **polarised** — they have **some covalent character** — whereas the bonds in sodium halides have almost no polarisation and very little covalent character.

See next page for more on polarisation of ionic bonds.

Polarisation of Ionic Bonds Leads to ***Covalent Character*** in Ionic Lattices

So, **magnesium** halides have more covalent character in their ionic bonds than sodium halides. Here's why...

1) In a sodium halide, e.g. NaCl, the **cation**, Na^+, has only a **small charge** (+1) so it can't really pull electrons towards itself — so the charge is distributed evenly around the ions (there's almost **no polarisation**).
2) This is pretty much what the simple **ionic model** looks like — that's why the theoretical calculations of lattice energy match the experimental ones so well for sodium halides.
3) However, the magnesium halides don't fit the ionic model quite so well, because charge isn't evenly distributed around the ions — the cation, Mg^{2+}, has a **bigger charge** (+2), so it can pull electrons towards itself a bit, polarising the bond.
4) In general, the greater the **charge density** of the **cation** (its charge compared to its volume) the poorer the match will be between experimental and theoretical values for lattice energy. (There's more on this on the next page.)

Lattice Energies & Polarisation of Ions

Small Cations Are Very Polarising

What normally happens in ionic compounds is that the **positive charge** on the **cation** attracts electrons towards it from the **anion** — this is **polarisation**.

A **cation** is just a positive ion, an **anion** is a negative ion, and an **onion** is the edible bulb of the Allium cepa plant.

1) **Small** cations with a **large charge** are **very polarising** because they have a **high charge density** — the positive charge is concentrated in the ion. So the cation can pull electrons towards itself.
2) **Large anions** are **polarised more easily** than small anions because their electrons are further away from the nucleus. So the electrons on large anions can be pulled away more easily towards cations.
3) If a compound contains a cation with a **high polarising ability** and an anion which is **easily polarised**, some of the anion's electron charge cloud will be dragged towards the positive cation.
4) If the compound is polarised enough, a partially **covalent bond** is formed.

How polarising a cation is depends on its charge density. The charge density is just the charge/volume ratio.

$$\text{Charge density} = \frac{\text{charge}}{\text{volume}}$$

Increasing the positive charge leads to more polarisation —

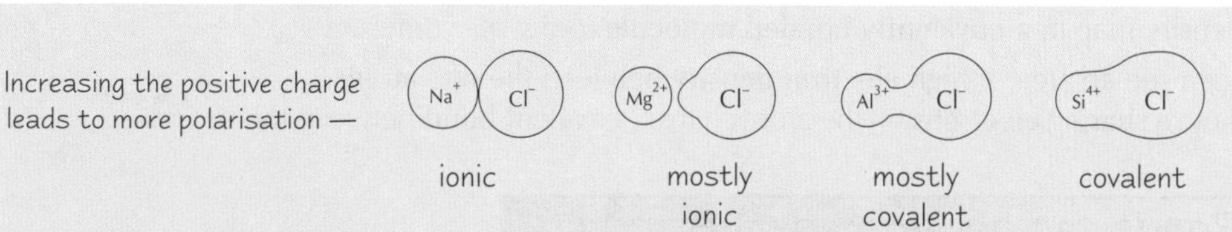

- The more an **ionic bond** is polarised, the more **covalent character** it gains.
- You can work out why some compounds have **weird properties** by thinking about how **polarised** the compound is.

Practice Questions

Q1 How can you tell from a Born-Haber cycle whether an ionic compound is significantly polarised?

Q2 What is a dipole?

Q3 What sort of cation is highly polarising? What sort of anion is easily polarised?

Exam Questions

Q1 Metal/non-metal compounds are usually ionic, yet solid aluminium chloride exhibits many covalent characteristics. Explain why. [4 marks]

Q2 Consider the following compounds:

$MgBr_2$ NaBr MgI_2

a) These compounds have differing degrees of covalent character in their bonds. Arrange the compounds in order of increasing covalent character, and explain your reasoning. [3 marks]

b) The theoretical lattice enthalpy of sodium iodide matches well with its experimental value but the theoretical lattice enthalpy of magnesium iodide does not match well with its experimental value.

Explain this difference. [2 marks]

Lattice Energy — it's why rabbits have so many babies...

Right then, you might be expected to interpret some data like that on page 42. And it's not enough to say, 'This lattice energy value's bigger than this one' — you have to explain what the pattern in the data shows. Remember, the closer the two lattice energy values, the better the 'purely ionic' model fits your compound — and the less polarised the bonds are.

Covalent Bonding

And now for covalent bonding — this is when atoms share electrons with one another so they've all got full outer shells.

Molecules are Groups of Atoms **Bonded** Together

1) Molecules are formed when **2 or more** atoms bond together — the atoms can be the **same** or **different**. Chlorine gas (Cl_2), carbon monoxide (CO), water (H_2O) and ethanol (C_2H_5OH) are all molecules.
2) Molecules are held together by **covalent bonds**, which are **strong**.

In covalent bonding, two atoms **share** electrons, so they've **both** got full outer shells of electrons. Both the positive nuclei are attracted **electrostatically** to the shared electrons.

E.g. two hydrogen atoms bond covalently to form a molecule of hydrogen.

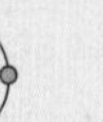

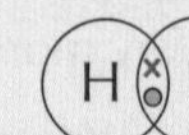

Covalent bonding happens between non-metals. Ionic bonding is between a metal and a non-metal.

Electron Density Maps Give Evidence for Covalent Bonding

On page 39 you saw what the **electron density map** of an ionic compound looks like. The electron density map of a **covalently bonded molecule** looks very different.

This time you can see an area of **high electron density** between the two atoms. It shows that they're **sharing electrons** — the atoms have a **covalent bond** between them.

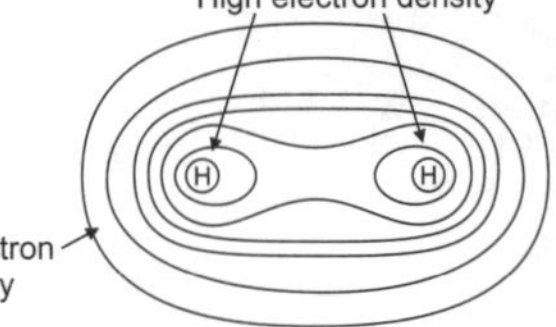

Covalent Bonds can be **Sigma (σ) Bonds...**

1) Hydrogen atoms each have an electron in an **s orbital** (have a look back at page 28 if you've forgotten what an s orbital is). When the hydrogen atoms form an H_2 molecule, the two s orbitals overlap to make a **σ bond** (sigma bond):
2) The two s orbitals overlap in a straight line — this gives the **highest possible electron density** between the two nuclei. This is a **single** covalent bond and is shown as a single line between the atoms, like this: H — H

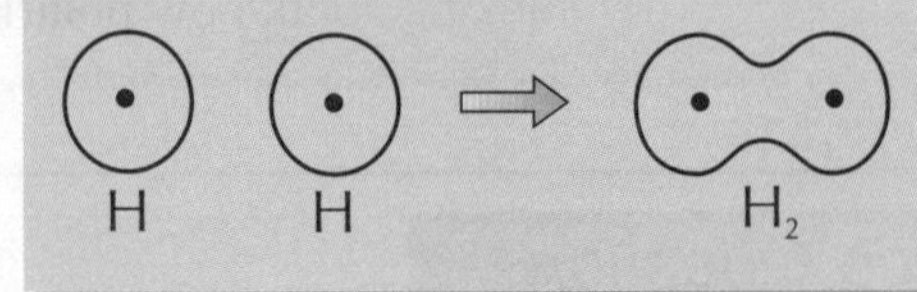

...or **Pi (π) Bonds**

1) A **π bond** is formed when two electrons in **p orbitals** overlap.
2) It's got **two parts** to it — one 'above' and one 'below' the molecular axis. This is because the π orbitals which overlap are **dumbbell shaped** (again, see page 28 if you're bewildered).
3) The π bond is **less tightly bound** to the two nuclei than the σ bond. This means π bonds are **weaker** than σ bonds and molecules with π bonds are **more reactive**.

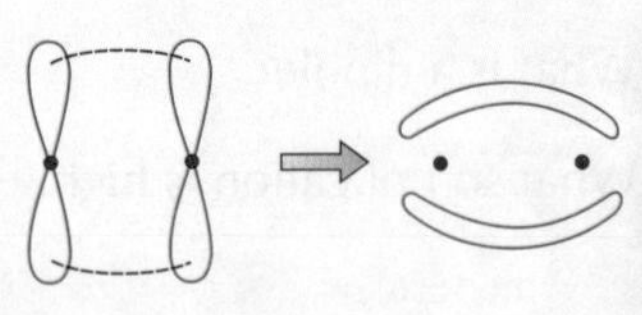

Atoms don't just form single bonds — **double** or even **triple covalent bonds** can form too. An example of a compound with a double bond is **ethene**, C_2H_4.

Its carbon-carbon **double** bond is drawn as C=C, but you're not really going to get little equals signs holding atoms together. The double bond is actually made up of a **σ bond** plus a **π bond**.

Ethene's π bond makes it a lot **more reactive** than ethane, which has only got σ bonds.

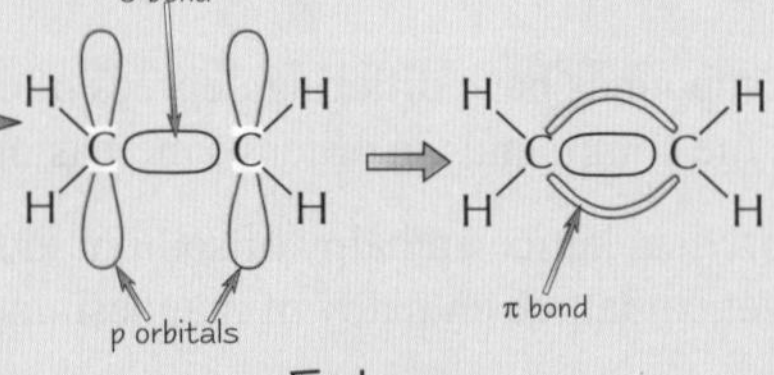

Ethene

You can also show ethene's bonding with a **dot-and-cross diagram**.

Covalent Bonding

Make sure you can Draw the Bonding in these Molecules

These diagrams don't show all the electrons in the molecules — just the ones in the **outer shells**:

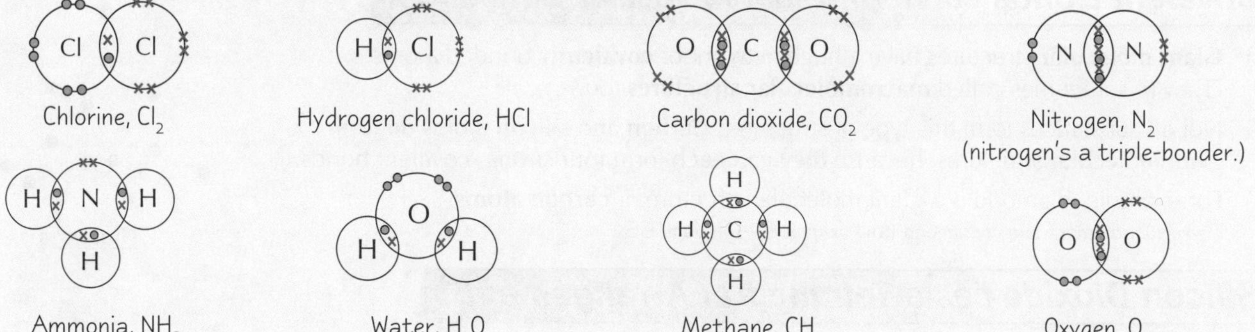

Dative Covalent Bonding is where Both Electrons come from One Atom

The **ammonium ion** (NH_4^+) is formed by dative (or coordinate) covalent bonding — it's an example the examiners love. It forms when the nitrogen atom in an ammonia molecule **donates a pair of electrons** to a proton (H^+) —

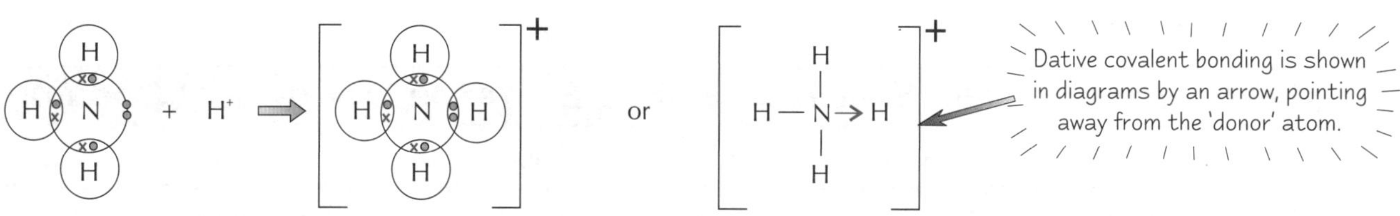

Practice Questions

Q1 a) When two or more atoms combine, they form a ____________.

b) In covalent bonding, electrons are ________ between atoms.

Q2 Draw a dot-and-cross diagram to show the arrangement of the outer electrons in a molecule of hydrogen chloride.

Q3 How does an electron density map provide evidence for the covalent bonding model?

Exam Questions

Q1 Methane has the molecular formula CH_4.

a) What type of bonding would you expect it to have? [1 mark]

b) Draw a dot-and-cross diagram showing the arrangement of <u>all</u> the electrons in a molecule of methane. [2 marks]

Q2 In terms of covalent bonds, explain why ethene is more reactive than ethane. [3 marks]

Q3 a) Draw a dot-and-cross diagram of the ammonia molecule (NH_3) showing the outer shell electrons only. [2 marks]

b) Draw a dot-and-cross diagram of the hydrogen chloride molecule (HCl) showing the outer shell electrons only. [2 marks]

c) Ammonia reacts with hydrogen chloride to form ammonium chloride. Draw a dot-and-cross diagram to show the bonding in ammonium chloride. [3 marks]

Steak and kidney — a great pie bond...

More pretty diagrams to learn here folks — practise till you get every single dot and cross in the right place. It's totally amazing to think of these titchy little atoms sorting themselves out so they've got full outer shells of electrons. Remember — covalent bonding happens between two non-metals, whereas ionic bonding happens between a metal and a non-metal.

Giant Covalent Structures & Metallic Bonding

Atoms can form giant structures as well as piddling little molecules — well...'giant' in molecular terms anyway. Compared to structures like the Eiffel Tower or even your granny's carriage clock, they're still unbelievably tiny.

Covalent Bonds *can form Giant Molecular Structures*

1) **Giant molecular** structures have a huge network of **covalently** bonded atoms. (They're sometimes called **macromolecular structures** too.)
2) Not all substances form this type of structure. **Carbon** and **silicon** atoms **do** form giant molecular structures, because they can each form four strong, covalent bonds.
3) For example, diamond is a giant molecular structure of **carbon atoms**.
There's loads more about diamond (and graphite) —on page 66.

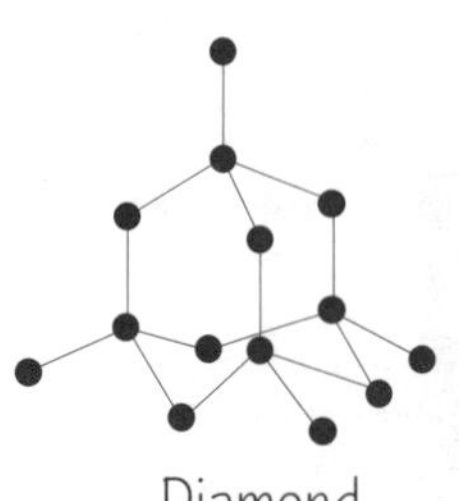
Diamond

Silicon Dioxide *has a* ***Tetrahedral Arrangement***

1) **Silicon** atoms form a giant covalent structure when they bond with **oxygen** atoms to form **silicon dioxide** — SiO_2. Silica is found as **quartz** or **sand** (sand's not pure, it's got lots of bits of other stuff in too).
2) Each silicon atom **covalently bonds** with **four oxygen atoms** in a **tetrahedral** arrangement to form a big **crystal lattice**.
3) Its structure **isn't** exactly the same as diamond's, because the oxygen atoms can only bond with **two silicon atoms**.

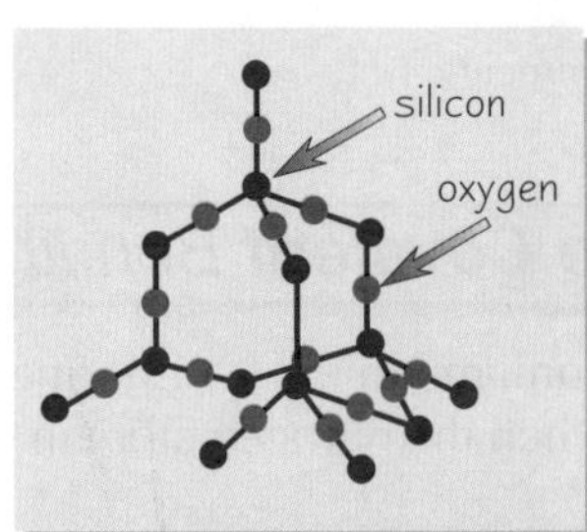

The ***Properties*** *of Giant Structures Provide* ***Evidence*** *for* ***Covalent Bonding***

All of these giant structures have some properties in common — thanks to their covalent bonds:

1) They are all **insoluble** in **polar solvents** like water, which shows that they **don't contain ions.**
2) They form **hard crystals** with very **high melting points**. This is down to their network of very strong covalent bonds.
3) They **don't conduct electricity**. All their bonding electrons are used to form **covalent bonds**, and they contain **no charged particles**. (The exception to this rule is graphite, which can conduct electricity because of **delocalised electrons** within its sheets of atoms.)

Metals *have Giant Structures Too*

Metal elements exist as **giant metallic lattice structures**.

1) The outermost shell of electrons of a metal atom is **delocalised** — the electrons are free to move about the metal. This leaves a **positive metal ion**, e.g. Na^+, Mg^{2+}, Al^{3+}.
2) The positive metal ions are **attracted** to the delocalised negative electrons. They form a lattice of closely packed positive ions in a **sea** of delocalised electrons — this is **metallic bonding**.

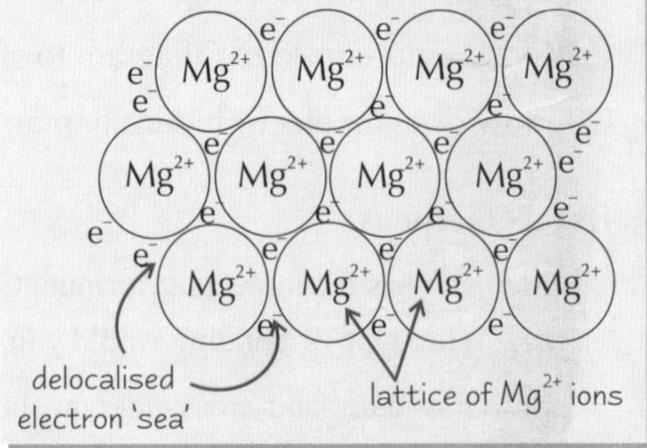

Metallic Bonding *Explains the* ***Properties*** *of Metals*

Metallic bonding explains why metals do what they do —

1) The **number of delocalised electrons per atom** affects the melting point. The **more** there are, the **stronger** the bonding will be and the **higher** the melting point. Mg^{2+} has **two** delocalised electrons per atom, so it's got a **higher melting point** than Na^+, which only has **one**.
2) The **size** of the metal ion also affects the melting point, because it affects the ion's **charge density** (charge to volume ratio). The higher the charge density, the stronger the bonding and the higher the melting point.
3) As there are **no bonds** holding specific ions together, the metal ions can slide over each other when the structure is pulled, so metals are **malleable** (a posh word for shapeable) and **ductile** (can be drawn into a wire).
4) The delocalised electrons can pass **kinetic energy** to each other, making metals **good thermal conductors**.
5) Metals are **good electrical conductors** because the **delocalised electrons** can carry a **current**.
6) Metals are **insoluble**, except in **liquid metals**, because of the **strength** of the metallic bonds.

Giant Covalent Structures & Metallic Bonding

Learn the **Properties** of the Main Substance Types

Make sure you know this stuff like the back of your spam —

Bonding	Examples	Melting and boiling points	Typical state at STP	Does solid conduct electricity?	Does liquid conduct electricity?	Is it soluble in water?
Ionic	NaCl $MgCl_2$	High	Solid	No (ions are held firmly in place)	Yes (ions are free to move)	Yes
Simple molecular* (covalent)	CO_2 I_2 H_2O	Low (have to overcome London forces or hydrogen bonds, not covalent bonds)	May be solid (like I_2), but usually liquid or gas (water is liquid because it has hydrogen bonds)	No	No	Depends on how polarised the molecule is
Giant molecular (covalent)	Diamond Graphite SiO_2	High	Solid	No (except graphite)	— (will generally sublime)	No
Metallic	Fe Mg Al	High	Solid	Yes (delocalised electrons)	Yes (delocalised electrons)	No

*To melt or boil simple covalent compounds you only have to overcome the weak intermolecular forces (London forces — see p70) that hold the molecules together. These are far weaker than ionic, covalent or metallic bonds. That's why their melting points are low.

Practice Questions

Q1 How does the model of metallic bonding explain why calcium has a higher melting point than potassium?

Q2 Explain why diamond is a non-conductor of electricity.

Q3 Why does silicon dioxide have a high melting point?

Q4 In metallic bonding, metal ____________ are held together by a "sea" of ____________ electrons. These electrons are free to move through the metal lattice, which explains why metals ____________ electricity even when solid.

Exam Questions

Q1 Illustrate with a suitable labelled diagram the structure of a typical metal and explain what is meant by metallic bonding. [3 marks]

Q2 The table below gives some data for the compound boron nitride, BN.

Melting point	Solubility in water	Electrical Conductivity of solid
3027 °C (sublimes)	insoluble	zero

Use this data to suggest the bond type and structure of boron nitride, and explain your answer. [4 marks]

Q3 a) With reference to the periodic table, place the following metals in order of increasing melting point:
aluminium, potassium, sodium [2 marks]

b) Explain your answer to part a) in terms of metallic bonding. [3 marks]

Carbon — it's a girl's best friend...

Remember, giant covalent structures are a different kettle of fish from simple covalent molecules (like CO_2, say). In simple covalent substances, separate molecules waft around with only weak intermolecular bonds between them. In a giant covalent structure, strong covalent bonds hold each atom in its place — and this explains their very different properties.

Organic Groups

Here it is, the final section of Unit 1, but it's a biggie. Organic chemistry is all about carbon compounds. Read on...

There are Loads of Ways of Representing Organic Compounds

TYPE OF FORMULA	WHAT IT SHOWS YOU	FORMULA FOR BUTAN-1-OL
General formula	An algebraic formula that can describe **any member** of a family of compounds.	$C_nH_{2n+1}OH$ (for all alcohols)
Empirical formula	The **simplest ratio** of atoms of each element in a compound (cancel the numbers down if possible). (So ethane, C_2H_6, has the empirical formula CH_3.)	$C_4H_{10}O$
Molecular formula	The **actual** number of atoms of each element in a molecule, with any **functional groups** indicated.	C_4H_9OH
Structural formula	Shows the atoms **carbon by carbon**, with the attached hydrogens and functional groups.	$CH_3CH_2CH_2CH_2OH$
Displayed formula	Shows how all the atoms are **arranged**, and all the bonds between them.	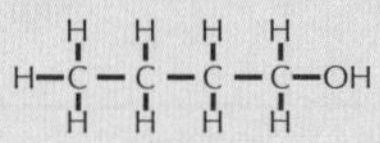 H–C–C–C–C–OH (each C with H above and below)
Skeletal formula	Shows the **bonds** of the carbon skeleton **only**, with any functional groups. The hydrogen and carbon atoms aren't shown. This is handy for drawing large complicated structures, like cyclic hydrocarbons.	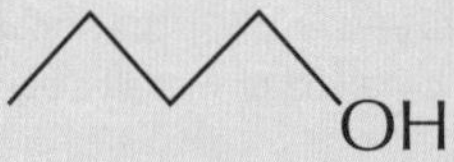 OH

A functional group is a reactive part of a molecule — it gives it most of its chemical properties.

Nomenclature is a Fancy Word for the Naming of Organic Compounds

You can name any organic compound using these **rules** of nomenclature.

1) Count the carbon atoms in the **longest continuous chain** — this gives you the stem:

Number of carbons	1	2	3	4	5	6	7	8	9	10
Stem	meth-	eth-	prop-	but-	pent-	hex-	hept-	oct-	non-	dec-

2) The **main functional group** of the molecule usually gives you the end of the name (the **suffix**) — see the table on the next page.

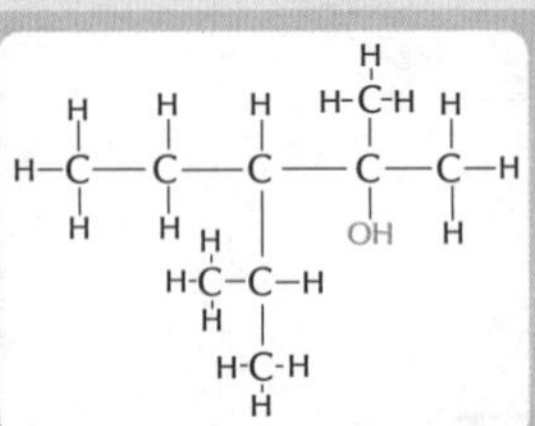

The longest chain is 5 carbons, so the stem is **pent-**

The main functional group in this example is **-OH**, so it's **pentanol**.

Don't forget — the longest carbon chain may be bent.

3) Number the carbons in the **longest** carbon chain so that the carbon with the main functional group attached has the lowest possible number.
   If there's more than one longest chain, pick the one with the **most side-chains**.
4) Write the carbon number that the functional group is on **before the suffix**.

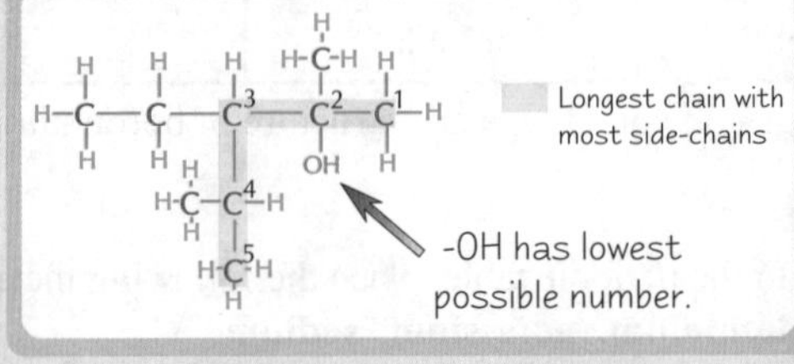

Longest chain with most side-chains

-OH has lowest possible number.

–OH is on carbon-2, so it's some sort of **"pentan-2-ol"**.

5) Any side-chains or less important functional groups are added as prefixes at the start of the name. Put them in **alphabetical** order, with the **number** of the carbon atom each is attached to.
6) If there's more than one **identical** side-chain or functional group, use **di-** (2), **tri-** (3) or **tetra-** (4) before that part of the name — but ignore this when working out the alphabetical order.

There's an ethyl group on carbon-3, and methyl groups on carbon-2 and carbon-4, so it's **3-ethyl-2,4-dimethylpentan-2-ol**

Organic Groups

Homologous Compounds have the Same General Formulas

1) A **homologous series** of organic compounds is a group of compounds that all have the **same general formula**.
2) Each member of a series has the **same functional group** but **differs by $-CH_2-$** in its carbon chain.

E.g. **Alcohols** have the general formula $C_nH_{2n+1}OH$. Here are the first six in the homologous series.

Methanol	Ethanol	Propanol	Butanol	Pentanol	Hexanol
CH_3OH	C_2H_5OH	C_3H_7OH	C_4H_9OH	$C_5H_{11}OH$	$C_6H_{13}OH$

You can see how each alcohol has **one carbon** and **two hydrogens** more than the one before.

3) You need to know about four of these series — alkanes, alkenes, halogenoalkanes and alcohols. Here are some examples of each:

Homologous series	Prefix or Suffix	Example
alkanes	-ane	propane: $CH_3CH_2CH_3$
branched alkanes	alkyl- , (-yl)	methylpropane: $CH_3CH(CH_3)CH_3$
alkenes	-ene	propene: $CH_3CH{=}CH_2$
halogenoalkanes	chloro- / bromo- / iodo-	chloroethane: CH_3CH_2Cl
alcohols	-ol	ethanol: CH_3CH_2OH

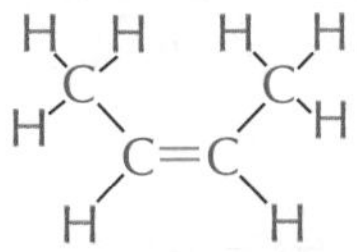

If the double bond in an alkene could go in more than one place, you have to say which carbon it starts on. This is but-2-ene.

Practice Questions

Q1 Explain the difference between molecular formulas and structural formulas.

Q2 In what order should prefixes be listed in the name of an organic compound?

Q3 What is a homologous series? Give four examples of homologous series.

Q4 What do you call an alcohol with three carbons?

Exam Questions

Q1 1-bromobutane is prepared from butan-1-ol in this reaction: $C_4H_9OH + NaBr + H_2SO_4 \rightarrow C_4H_9Br + NaHSO_4 + H_2O$

a) Draw the displayed formulae for butan-1-ol and 1-bromobutane. [2 marks]

b) What is the functional group in butan-1-ol and why is it necessary to state its position on the carbon chain? [2 marks]

c) Name the homologous series that butan-1-ol is a member of. [1 mark]

Q2 a) Name the following molecules.

(i) [2 marks]

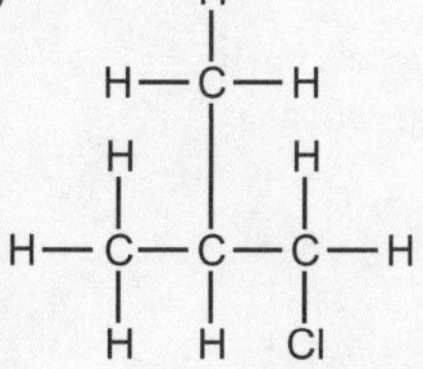

(ii) [2 marks]

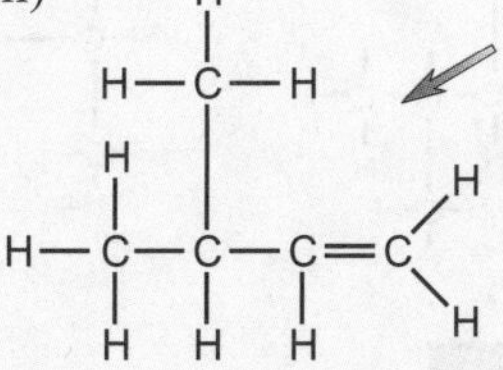

The double bond is the most important functional group, so give it the lowest number.

(iii) [2 marks]

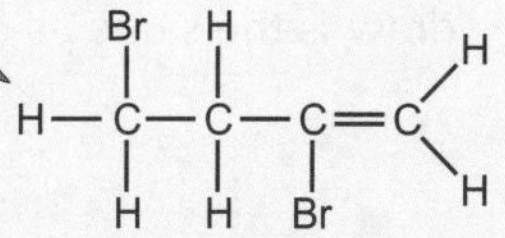

b) (i) Write down the molecular formula for 3-ethylpentane. [1 mark]

(ii) Write down the structural formula for this molecule. [1 mark]

It's as easy as 1,2,3-trichloropentan-2-ol...

The best thing to do now is find some random organic compounds and work out their names using the rules. Then have a bash at it the other way around — read the name and draw the compound. It might seem a wee bit tedious now, but come the exam, you'll be thanking me. Doing the exam questions will give you some good practice too.

Alkanes and Structural Isomerism

I'm an alkane and I'm OK — I sleep all night and I work all day...

Alkanes are **Saturated Hydrocarbons**

1) Alkanes have the **general formula C_nH_{2n+2}**. They've only got **carbon** and **hydrogen** atoms, so they're **hydrocarbons**.
2) Every carbon atom in an alkane has **four single bonds** with other atoms. It's **impossible** for carbon to make more than four bonds, so alkanes are **saturated**. Here are a few examples of alkanes:

Methane — Ethane — Propane

3) You get **cycloalkanes** too. They have a ring of carbon atoms with two hydrogens attached to each carbon.
4) Cycloalkanes have a **different general formula** from that of normal alkanes (C_nH_{2n}, assuming they have only one ring), but they're still **saturated**.

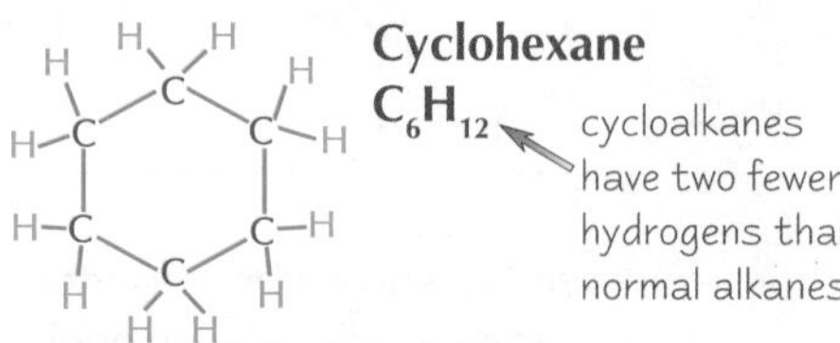

Structural Isomers Have ***Different Arrangements*** of the Same ***Atoms***

In structural isomers the atoms are **connected** in different ways. But they still have the **same molecular formula**. Alkanes can have **chain isomers**. (There are other types of structural isomer, but don't worry about them till Unit 2.)

CHAIN ISOMERS

Chain isomers have different arrangements of the **carbon skeleton**. Some are **straight chains** and others **branched** in different ways.

Both of these compounds have the same **molecular formula** — C_4H_{10}.
But you can show the difference by writing their **structural formulas**:

butane — structural formula: $CH_3CH_2CH_2CH_3$

methylpropane — structural formula: $CH_3CH(CH_3)CH_3$

Don't be Fooled — What Looks Like an Isomer Might ***Not*** Be

Atoms can rotate as much as they like around single **C–C bonds**. Remember this when you work out structural isomers — sometimes what looks like an isomer, isn't.

For example, there are only **three** chain isomers of C_5H_{12}...

pentane — methylbutane — 2,2-dimethylpropane

...because everything else you might draw is the same as one of these. E.g.

These are both methylbutane.

Alkanes Burn ***Completely*** in Oxygen

1) If you burn (**oxidise**) alkanes with **oxygen**, you'll get **carbon dioxide** and water — this is a **combustion reaction**.

Here's the equation for the combustion of propane — $C_3H_{8(g)} + 5O_{2(g)} \rightarrow 3CO_{2(g)} + 4H_2O_{(g)}$

2) Combustion reactions happen between **gases**, so liquid alkanes have to be **vaporised** first.
Smaller alkanes turn into **gases** more easily (they're more **volatile**), so they'll **burn** more easily too.

Alkanes and Structural Isomerism

Halogens React with Alkanes, Forming Halogenoalkanes

A hydrogen atom is **substituted** (replaced) by chlorine or bromine in a **photochemical** reaction (a reaction started by UV radiation). This is a **free-radical substitution reaction**.

Free radicals are particles with an unpaired electron, written Cl• or CH_3• — you get them when bonds split equally.

For example, **chlorine** and **methane** react with a bit of a bang to form **chloromethane**: $CH_4 + Cl_2 \xrightarrow{UV} CH_3Cl + HCl$

The **reaction mechanism** has three stages:

Initiation reactions — free radicals are produced.

1) Sunlight provides enough energy to break the Cl-Cl bond — this is **photodissociation**.
2) The bond splits **equally** and each atom gets to keep one electron — **homolytic fission**. The atom becomes a highly reactive **free radical**, Cl·, because of its **unpaired electron**.

The movement of the electrons is shown by curly half arrows.

Cl–Cl ⟶ Cl· + Cl·

Propagation reactions — free radicals are used up and created in a chain reaction.

1) Cl· attacks a **methane** molecule: Cl· + CH_4 → CH_3· + HCl
2) The new **methyl free radical**, CH_3·, can attack another Cl_2 molecule: CH_3· + Cl_2 → CH_3Cl + Cl·
3) The new Cl· can attack **another** CH_4 molecule, and so on, until all the Cl_2 or CH_4 molecules are wiped out.

Termination reactions — free radicals are mopped up.

1) If two free radicals join together, they make a **stable molecule**.
2) There are **heaps** of possible termination reactions. Here's a couple of them to give you the idea:

Cl· + CH_3· → CH_3Cl

CH_3· + CH_3· → C_2H_6

Some products formed will be trace impurities in the final sample. The presence of ethane is evidence for the mechanism.

What happens now **depends** on whether there's too much **chlorine** or too much **methane**:

1) If the **chlorine's** in excess, Cl· free radicals will start attacking chloromethane, producing **dichloromethane** CH_2Cl_2, **trichloromethane** $CHCl_3$, and **tetrachloromethane** CCl_4.
2) But if the **methane's** in excess, then the product will mostly be **chloromethane**.

Practice Questions

Q1 What's the general formula for alkanes?

Q2 What are isomers?

Q3 What's a free radical?

Q4 What's homolytic fission?

Exam Questions

Q1 There are five different compounds that all have the molecular formula C_6H_{14}.

a) Draw the displayed formula of each compound, and give its full name. [10 marks]

b) These compounds are part of a homologous series. Name the series. [1 mark]

Q2 The alkane ethane is a saturated hydrocarbon. It is mostly unreactive, but will react with oxygen in a combustion reaction and bromine in a photochemical reaction.

a) What is a saturated hydrocarbon? [2 marks]

b) Write a balanced equation for the complete combustion of ethane. [2 marks]

c) Write an overall equation and outline the mechanism for the photochemical reaction of bromine with ethane. Assume ethane is in excess. What type of mechanism is it? [8 marks]

This page is like...totally radical, man...

Mechanisms can be an absolute pain in the bum to learn, but unfortunately reactions are what Chemistry's all about. If you don't like it, you should have taken art — no mechanisms in that, just pretty pictures. Ah well, there's no going back now. You've just got to sit down and learn the stuff. Keep hacking away at it, till you know it all off by heart.

Petroleum

Petroleum is just a poncy word for crude oil — the black, yukky stuff they get out of the ground.

Crude Oil is a Mixture of Hydrocarbons

1) Petroleum or crude oil is mostly **alkanes**. They range from **smallish alkanes**, like pentane, to **massive alkanes** with more than 50 carbons.
2) Crude oil isn't very useful as it is, but you can **separate** it into more useful bits (or **fractions**) by **fractional distillation**.

Here's how fractional distillation works — don't try this at home.

1) First, the crude oil is **vaporised** at about 350 °C.
2) The vaporised crude oil goes into the **fractionating column** and rises up through the trays. The largest hydrocarbons don't **vaporise** at all, because their boiling points are too high — they just run to the bottom and form a gooey **residue**.
3) As the crude oil vapour goes up the fractionating column, it gets **cooler**. Because of the different chain lengths, each fraction **condenses** at a different temperature. The fractions are **drawn off** at different levels in the column.
4) The hydrocarbons with the **lowest boiling points** don't condense. They're drawn off as **gases** at the top of the column.

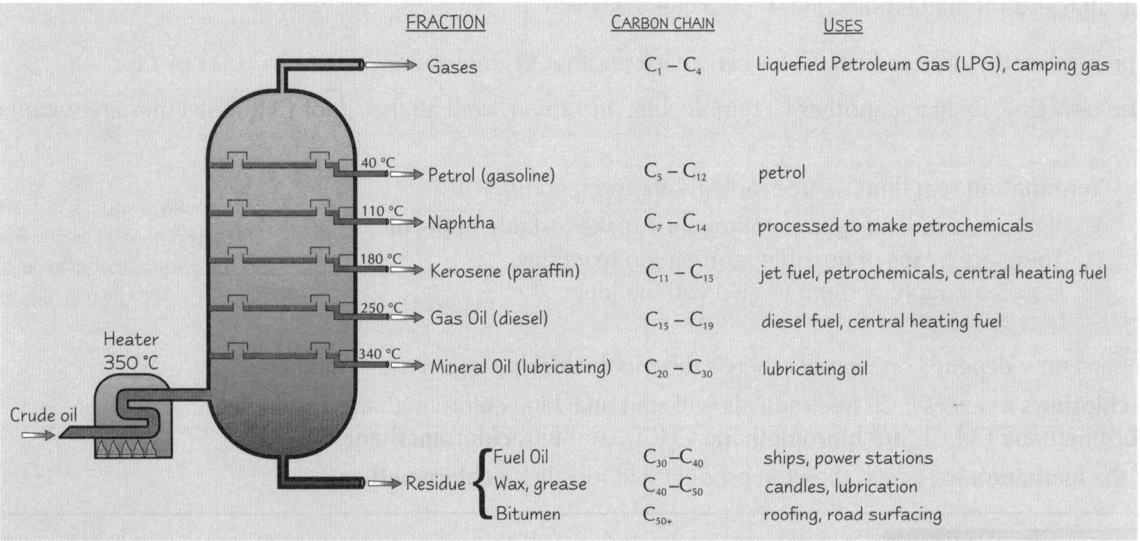

Heavy Fractions can be 'Cracked' to Make Smaller Molecules

1) People want loads of the **light** fractions, like petrol and naphtha. They don't want so much of the **heavier** stuff like bitumen though. Stuff that's in high demand is much more **valuable** than the stuff that isn't.
2) To meet this demand, the less popular heavier fractions are **cracked**. Cracking is **breaking** long-chain alkanes into **smaller** hydrocarbons (which can include alkenes). It involves breaking the **C–C bonds**.

   You could crack **decane** like this —

$$\underset{\text{decane}}{C_{10}H_{22}} \rightarrow \underset{\text{ethene}}{C_2H_4} + \underset{\text{octane}}{C_8H_{18}}$$

There are **two types** of **cracking** you need to know about:

THERMAL CRACKING

- It takes place at **high temperature** (up to 1000 °C) and **high pressure** (up to 70 atm).
- It produces a lot of **alkenes**.
- These **alkenes** are used to make heaps of valuable products, like **polymers**. A good example is **poly(ethene)**, which is made from ethene (have a squiz at page 62 for more on polymers).

CATALYTIC CRACKING

- This makes mostly **motor fuels** and **aromatic** hydrocarbons (see page 56).
- It uses something called a **zeolite catalyst**, at a **slight pressure** and **high temperature** (about 450 °C).
- Using a catalyst **cuts costs**, because the reaction can be done at a **lower** temperature and pressure. The catalyst also **speeds** up the reaction, and time is money and all that.

Petroleum

Fuels Contain a **Mixture** of **Types of Alkane**

1) Most people's cars run on petrol or diesel, both of which contain a mixture of alkanes (as well as other hydrocarbons, impurities and additives).
2) Some of the alkanes in petrol are **straight-chain** alkanes, e.g. hexane — $CH_3CH_2CH_2CH_2CH_2CH_3$.
3) Petrol also contains some shorter, **branched-chain** alkanes, e.g. 2,3-dimethylbutane — $CH_3CH(CH_3)CH(CH_3)CH_3$, **cycloalkanes** and **aromatic hydrocarbons**. These types of alkane makes the fuel burn **more efficiently**.
4) **Catalytic cracking** produces aromatic hydrocarbons. Straight-chain alkanes can also be **reformed** to make cycloalkanes and more aromatic hydrocarbons

Straight-chain alkanes tend to explode of their own accord when the fuel/air mixture in the engine is compressed. This causes 'knocking' and makes combustion inefficient.

Alkanes can be **Reformed** into **Cycloalkanes** and **Aromatic Hydrocarbons**

Reforming converts **alkanes** into **aromatic hydrocarbons** (arenes — see page 56).
It uses a catalyst (e.g. platinum stuck on aluminium oxide).

$CH_3CH_2CH_2CH_2CH_2CH_3$ **hexane** $\xrightarrow{Pt}$ **cyclohexane** $+ H_2$ $\longrightarrow$ **benzene** $+ 3H_2$

Practice Questions

Q1 How does fractional distillation work?

Q2 What is cracking?

Q3 Why is reforming used?

Exam Question

Q1 Crude oil is a source of fuels and petrochemicals. It's vaporised and separated into fractions using fractional distillation.

a) Some heavier fractions are processed using cracking.

(i) Explain why cracking is carried out. [2 marks]

(ii) Write a possible equation for the cracking of dodecane, $C_{12}H_{26}$. [1 mark]

b) Some hydrocarbons present in petrol are processed using reforming.

(i) Name two types of compound that are produced by reforming. [2 marks]

(ii) What effect do these compounds have on the petrol's performance? [1 mark]

c) Pentane is often converted into its isomers, which are then added to petrol. Draw and name two of these isomers. [4 marks]

Crude oil — not the kind of oil you could take home to meet your mother...

This ain't the most exciting page in the history of the known universe. Although in a galaxy far, far away there may be lots of pages on even more boring topics. But, that's neither here nor there, cos you've got to learn the stuff anyway. Get fractional distillation and cracking straight in your brain and make sure you know why people bother to do it.

Fuels and Climate Change

In the olden days, the River Thames sometimes froze over in winter and people set up 'Frost Fairs' on the ice — stalls selling tat, people drinking too much mulled wine and pigs being roasted on spits (yep, I wondered about the whole fire-hot-melty thing too). This is all true (I looked it up) but it probably won't be happening again for a bit.

Alkanes are Useful Fuels...

When you **burn** an alkane in plenty of air, you end up with **carbon dioxide** and **water** — see p50.
It's an **exothermic** combustion reaction. Alkanes make great fuels, because burning them releases lots of energy.

Here's a few uses of alkane fuels:

1) Methane's used for **central heating** and **cooking** in homes.
2) Alkanes with 5-12 carbon atoms are used in **petrol**.
3) Kerosene is used as **jet fuel**. Its alkanes have 11-15 carbon atoms.
4) **Diesel** is made of a mixture of alkanes with 15-19 carbon atoms.

...But They Produce Harmful Emissions...

We generate most of our **electricity** by burning **fossil fuels** (coal, oil and natural gas) in **power stations**. We also use loads and loads of fossil fuels for **transport** and **heating**. Burning all these fossil fuels causes a lot of **pollution**.

Sulfur oxides

Power stations add loads of sulfur oxides, including sulfur dioxide, to the air. **Scrubbers** are used to reduce the sulfur dioxide emissions, but they don't get rid of them all. Sulfur oxides are **really nasty**:

1) They're **poisonous** and cause problems for people with asthma.
2) Sulfur oxides dissolve in moisture in the air, and are converted into **sulfuric acid**, which causes **acid rain**. Acid rain makes **lakes** and **rivers** acidic, which kills fish and other aquatic life. It also kills trees and damages buildings.

Carbon monoxide and hydrocarbon particles

Sometimes there isn't enough **oxygen** in an engine for the fuel to burn completely. This is **incomplete combustion**.

1) When this happens, carbon dioxide and water aren't the only things produced. **Carbon monoxide**, a poisonous gas, is also produced.
2) Sometimes, some of the fuel in an engine comes out of the exhaust pipe without burning at all. These **unburned hydrocarbon** particles escape into the air. Some of the particles are really fine, and cause health problems when they're breathed in and reach the lungs.

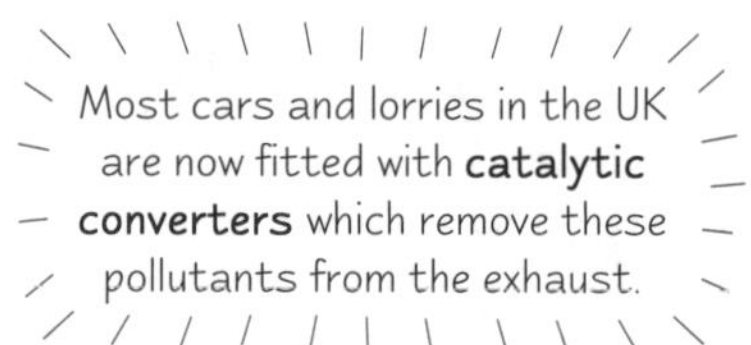

Nitrogen oxides

Vehicle engines also make **oxides of nitrogen** (NO_x).

1) These add to the **acid rain** problem.
2) Nitrogen dioxide, NO_2, causes **breathing problems**.
3) In **sunlight**, nitrogen dioxide reacts to produce ground-level **ozone**, another air pollutant.

...Including Greenhouse Gases

1) **Carbon dioxide is a greenhouse gas** (as are water vapour and methane).
2) Here's what greenhouse gases do:

   The Earth absorbs **radiation** from the Sun, and then it re-emits it as **infra-red radiation**. The greenhouse gases in the atmosphere **absorb** this radiation, then **re-emit** some of it towards Earth.

   In general, this is good, because it keeps the planet **warm enough** to support life, including us.

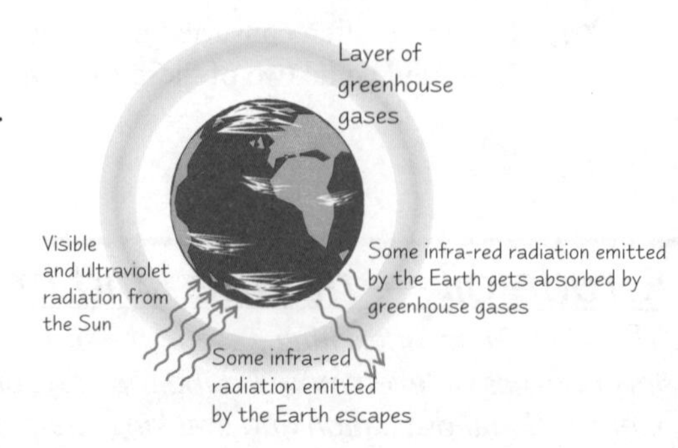

Fuels and Climate Change

Global Warming Will Probably Cause Significant *Climate Change*

So the **greenhouse effect** is what keeps the Earth nice and cosy. But you can have too much of a good thing — and **too much greenhouse effect** is causing **global warming**, which is probably a very **bad thing**.

1) Over the last 200 years or so, the **concentration** of **greenhouse gases** in the atmosphere has been **rising.**
2) That's mainly because humans have been putting **carbon dioxide into** the atmosphere faster than it's being **removed** (by photosynthesis in plants). Most of this extra carbon dioxide comes from **burning fossil fuels**.
3) And the extra CO_2 enhances the greenhouse effect, and is causing the Earth to **warm up** — global warming.
4) The vast majority of scientists working on this subject agree that global warming is causing **huge** problems:

- The **polar ice caps** are **melting**.
- **Sea levels** are rising (partly because of the ice melting, partly because warming the water in the oceans makes it expand) — this will cause widespread **flooding**.
- The world's **climate** is thought to be **changing** already — and climate change is likely to cause all kinds of problems, including severe shortages of food and fresh water.

Burning *Fossil Fuels* Just Isn't *Sustainable...*

1) As if pollution and global warming aren't bad enough, fossil fuels are **non-renewable** — we're using them up much more quickly than they can be replaced. (It takes millions of years for fossil fuels to form, deep under the seabed.) So one day, fossil fuels will just **run out** (and before that they'll become ridiculously **expensive**).
2) All in all, burning fossil fuels at the current rate just **isn't sustainable** — if we carry on like this the next couple of generations will be **in the soup**... a rather hot, dirty soup.
3) With all this in mind, scientists are trying to develop **alternative fuels** — fuels that don't produce nasty emissions, don't contribute to global warming, and are **renewable**. See pages 110 and 114 for more on alternative fuels.

Practice Questions

Q1 How does the combustion of hydrocarbon fuels contribute to the greenhouse effect?

Q2 Why is the production of sulfur dioxide harmful for the environment?

Q3 Give two problems associated with the gas NO_2.

Q4 What does 'non-renewable' mean?

Exam Questions

Q1 One of the components in petrol is the alkane pentane (C_5H_{12}).

a) Write a balanced equation for the complete combustion of pentane. [2 marks]

b) Explain how both products in the reaction in a) can contribute to the greenhouse effect. [3 marks]

c) Explain how the incomplete combustion of pentane could cause serious health problems. [2 marks]

Q2 Read the following:

Our current use of fossil fuels is not sustainable for two main reasons: because fossils fuels are a finite resource, and because carbon dioxide is emitted which is having a negative impact on the planet.

a) Explain what is meant by the term 'finite resource'. [1 mark]

b) Explain, in detail, why carbon dioxide emissions are 'having a negative impact on our planet'. [4 marks]

Fossil Fuels — they're bad for us but we still want them...

It's a dirty business, burning fossil fuels. You need to know about the various types of pollutants they release (sulfur oxides, nitrogen oxides, carbon monoxide, hydrocarbon particles) and what damage they do. And then of course there's greenhouse gases and climate change. Yup, if we could start again with industrialisation, we might do it differently.

Alkenes, Hazards and Risks

I'll warn you now — some of this stuff gets a bit heavy — but stick with it, cos it's pretty important.

Alkenes are **Unsaturated Hydrocarbons**

1) Alkenes have the **general formula C_nH_{2n}**. They're just made of carbon and hydrogen atoms, so they're **hydrocarbons**.
2) Alkene molecules **all** have at least one **C=C double covalent bond**. Molecules with C=C double bonds are **unsaturated** because they can make more bonds with extra atoms in **addition** reactions.

   Here are a few pretty diagrams of **alkenes**:

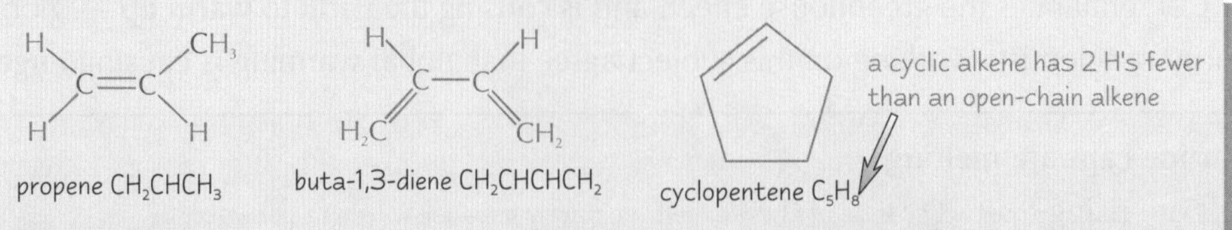

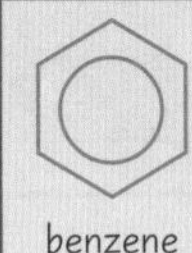
benzene

Benzene (C_6H_6) is like a **cyclic alkene** with 6 carbons and 3 double bonds. It's more **stable** (less reactive) than you'd expect though, because the double bond electrons are **delocalised** around the carbon ring. That's why its symbol has a **circle** in it.

3) Compounds with **benzene ring structures** are called **arenes**, or **aromatic compounds**. All other organic compounds (e.g. alkanes and alkenes) are called **aliphatic compounds**.

Alkenes are **Much More Reactive** than Alkanes

1) Each **double bond** in an alkene is made up of a **σ bond and a π bond** (see page 44). It's a bit like a hot dog — the **π bond** is the bun and the **σ bond** is sandwiched in the middle like the sausage.
2) Because there are two pairs of electrons in the bond, the C=C double bond has a really **high electron density**. This makes alkenes pretty reactive.
3) Another reason for the high reactivity is that the **π bond** sticks out above and below the rest of the molecule. So, the **π bond** is likely to be attacked by **electrophiles** (see below).
4) As the double bond is so **reactive**, alkenes are handy **starting points** for making other organic compounds and for making **petrochemicals**.

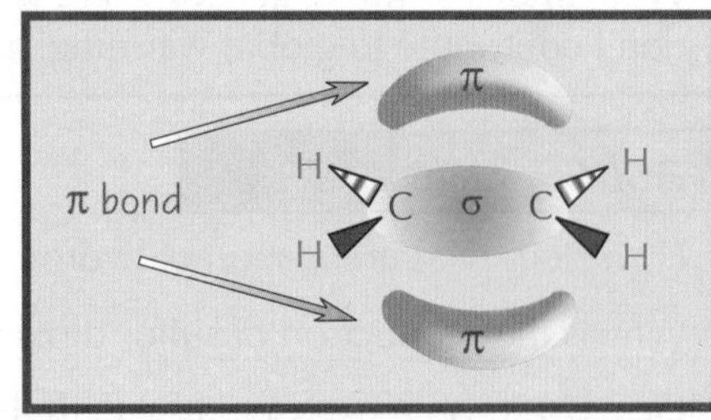

Adding **Hydrogen** to C=C Bonds Produces **Alkanes**

1) Ethene will react with **hydrogen** gas to produce ethane. It needs a **nickel catalyst** and a temperature of **150 °C** though.

$$H_2C{=}CH_2 + H_2 \xrightarrow[150\,°C]{Ni} CH_3CH_3$$

2) Similar reactions happen with other alkenes, e.g. propene reacts to produce propane.

Electrophilic Addition Reactions Happen to Alkenes

Electrophilic addition reactions of alkenes aren't too complicated.

1) The **double bond** opens up, and another atom is **added** to each of its carbons.
2) Addition reactions* happen because the double bond has got plenty of electrons and is easily attacked by an **electrophile**.
3) The double bond is also **nucleophilic** — it's attracted to places that don't have enough **electrons**.

*There's loads about electrophilic addition reactions on page 58 — this bit was just to get you all excited.

Electrophiles are **electron-pair acceptors** — they're usually a bit short of electrons, so they're **attracted** to areas where there's lots of them about.
Here are a couple of examples of electrophiles:

- **Positively charged ions**, like H^+, NO_2^+.
- **Polar molecules** — the δ+ atom is attracted to places with lots of electrons

One end of a polar molecule has a slight positive charge (δ+), the other end has a slight negative charge (δ–). See page 68 for more about polar molecules.

Alkenes, Hazards and Risks

I bet you're itching to get into the lab and do some electrophilic addition reactions. (I know, you feel ashamed, but it's completely natural.) But before you act on those chemistry-related urges, you should think about doing things safely.

Using Organic Chemicals can be **Hazardous**

You should already know the chemical **hazard symbols**. The ones below are common hazards of **organic chemicals**.

Most organic chemicals are flammable. Some (like ethanol and methane) are **highly flammable**.

Irritant: e.g. propan-1-ol, pentane. These chemicals can irritate or blister your skin, but won't cause permanent damage.

Toxic: e.g. methanol, chloroethene. Really nasty stuff — these chemicals can kill you if you swallow or inhale them, or sometimes even if you get them on your skin.

You won't see the '**explosive**' symbol in the lab much, but highly flammable gases can cause explosions if they're released in the air.

Harmful: e.g. butan-1-ol, 1-chloropropane. Pretty nasty stuff — harmful chemicals can damage your health, but aren't as dangerous as toxic ones.

This symbol means **dangerous for the environment**. You'll find it on chemicals that can cause very serious environmental damage (e.g. hexane).

You need to know the difference between **hazard** and **risk**:

A **hazard** is anything that can cause **harm**.
Risk is the **chance** that what you're doing will cause harm.

A **Risk Assessment** can Help to Make Lab Work **Safer**

1) Before you do **anything** with chemicals in a lab, you should do a **risk assessment**. A risk assessment looks at the **hazards** of all the **reactants**, **products** and **procedures** involved in an experiment and considers how to make the risks from them as **small as possible**. You can **reduce risks** by:
   - working on a smaller scale
   - taking appropriate precautions — eye protection, plastic gloves, fume cupboards, etc.
   - using different, safer chemicals or lower concentrations if possible
2) It's impossible to **completely** get rid of all risk. But the point of doing a risk assessment is to systematically think about ways to **minimise the risks**. Despite all the hazardous chemicals, science labs are actually pretty safe. If you follow all the rules, you're less likely to get injured in a chemistry lesson than you are travelling in a car.

Practice Questions

Q1 What's the general formula for an alkene?

Q2 What is an electrophile? Why do alkenes react with electrophiles?

Q3 Write an equation for the hydrogenation of ethene.

Exam Questions

Q1 The alkene myrcene can be isolated from bay leaves.
The structural formula of myrcene is $(CH_3)_2CCHCH_2CH_2C(CH_2)CHCH_2$.

a) What is its molecular formula? [1 mark]

b) How many double bonds does myrcene contain, assuming there are no cyclic groups in its structure? [1 mark]

c) Draw the displayed formula of myrcene. [2 marks]

Q2 The preparation of an organic compound involves chemicals with the hazard signs shown below.

Suggest two sensible ways to reduce the risks involved, other than wearing protective clothing such as goggles, laboratory overalls and plastic gloves. [2 marks]

Double bond hot dog — Brosnan's the bun, Dalton's the dog...

What ON EARTH are you talking about, man? Search me — I'm just here to tell you about Organic Chemistry. Alkenes are really important. Make sure you understand why they're so much more reactive than alkanes. It's all to do with the double bond. So get that Double Bond Hot Dog thingy learned. And remember — Connery's the ketchup, Moore's the mustard...

Reactions of Alkenes

Alkenes do loads of weird and wacky stuff — and there's a fair bit of it on these two pages.

Alkenes undergo Electrophilic Addition with Halogens...

On the last page I mentioned **electrophilic addition reactions**, and why they happen to alkenes. Well, here's one — if you mix an alkene and bromine, the bromine adds across the double bond to form a **dibromoalkane**. Here's the mechanism...

Other halogens, like chlorine and iodine, do this with alkenes too.

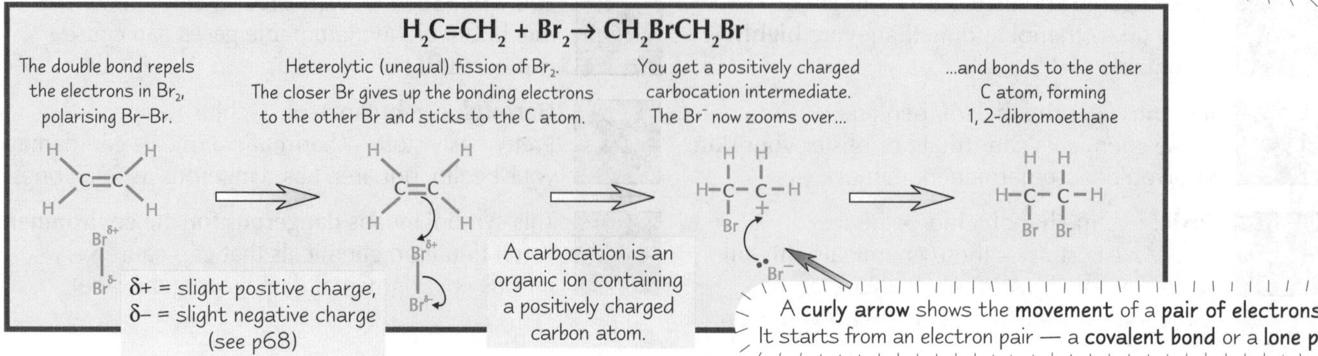

If you add some Cl^- ions to an ethene and bromine mixture, you'll also get some CH_2BrCH_2Cl.
This is evidence for the suggested mechanism — once the ethene has reacted with bromine to form the **carbocation** (**positively** charged), it can react with **either** another **bromide ion or** a **chloride ion** (**negatively** charged).

...and with Hydrogen Halides

Alkenes also undergo **electrophilic addition** reactions with hydrogen halides.
For example, ethene reacts with hydrogen bromide, HBr, to form **bromoethane**. Here's the mechanism:

HBr is polar because the electrons are more attracted to the Br than to the H, so they have a difference in charge. See page 68.

H–Br is a polar molecule. The δ+ H atom is attracted to the double bond electrons.

Heterolytic (unequal) fission of HBr. The H atom gives up the bonding electrons to the Br atom and accepts an electron pair from the double bond.

As before, the Br^- zooms over to the carbocation...

...and bonds to the other C atom, forming bromoethane

Adding Hydrogen Halides to Unsymmetrical Alkenes Forms Two Products

1) When HBr adds to a symmetrical molecule like ethene (as above), there's only one possible product.
2) But if a hydrogen halide adds to an **unsymmetrical** alkene, like propene, there are two possible products. The amount of each product formed depends on how **stable** the **carbocation** formed in the middle of the reaction is.
3) Carbocations with more **alkyl groups** are more stable, because the alkyl groups feed **electrons** towards the positive charge. The **more stable carbocation** is much more likely to form.

Alkyl groups are alkanes with a hydrogen removed, e.g. methyl, CH_3^-

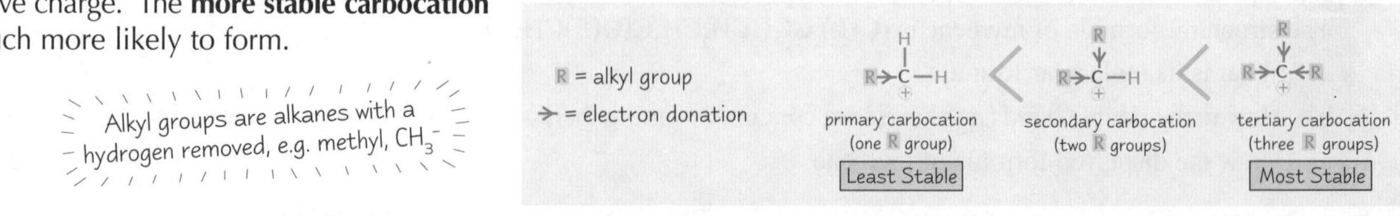

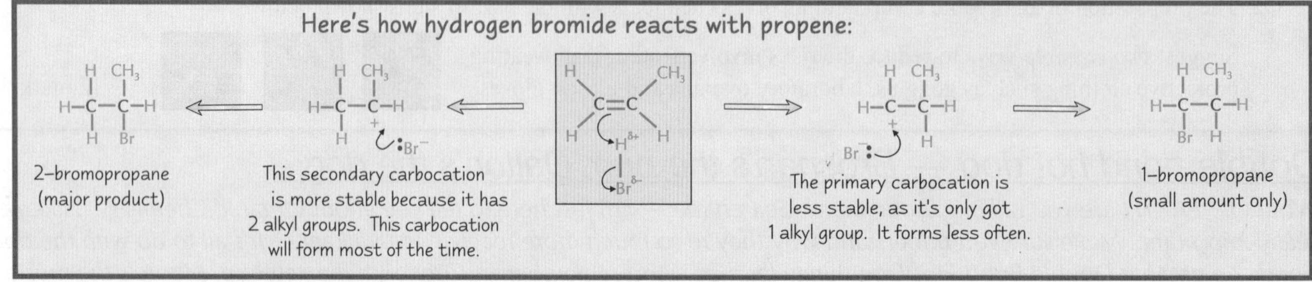

Reactions of Alkenes

Use **Bromine Water** to Test for C=C Double Bonds

1) When you shake an alkene with **orange bromine water**, the solution **decolourises**. You can use this reaction as a **test** for **C=C double bonds**.

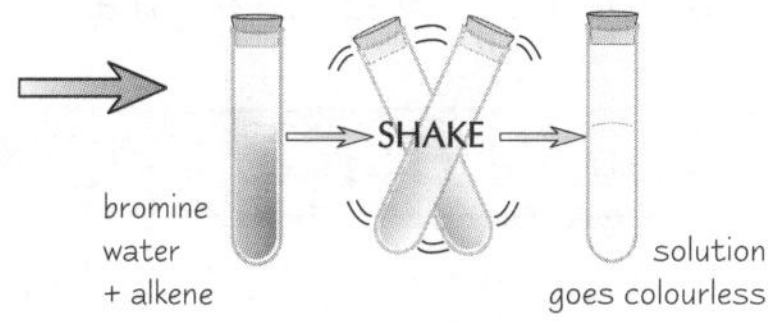

2) It'd be nice if this was a simple case of bromine molecules adding across the double bonds to make colourless dibromoalkanes, as on the previous page. Sadly, that's not quite the whole story...
3) Bromine water is a **dilute solution** — it contains more **water molecules** than bromine molecules. The **carbocation** is more likely to react with $\mathbf{H_2O}$ than Br^- — so an **OH** group adds to the second carbon rather than another Br.
4) This means that the product of the reaction between an alkene and bromine water is mostly **bromoalcohol**.

H H
H–C–C–H
Br OH
bromoethanol

Alkenes are **Oxidised** by **Acidified Potassium Manganate(VII)**

1) If you shake an alkene with **acidified potassium manganate(VII)**, the **purple** solution is **decolourised**. You've **oxidised** the alkene and made a diol (an alcohol with two -OH groups).
2) For example, here's how **ethene** reacts with acidified potassium manganate(VII):

[O] is often used to show an oxidising agent in an organic equation.

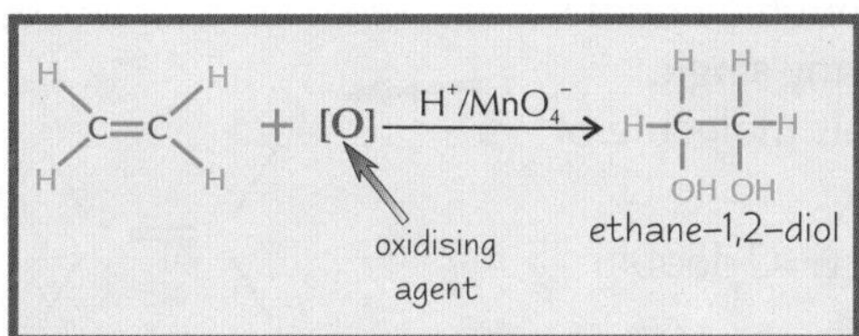

3) This reaction is another useful **test** for a double C=C bond.

Practice Questions

Q1 What product do you get when you react an alkane with a halogen?

Q2 In a mechanism diagram, what does a curly arrow show?

Q3 Which is the most stable kind of carbocation?

Q4 What reaction happens when you shake acidified potassium manganate(VII) solution with an alkene?

Exam Question

Q1 But-1-ene is an alkene. Alkenes contain at least one C=C double bond.

a) Describe how bromine water can be used to test for C=C double bonds. [2 marks]

b) What organic product is produced in this test? [1 mark]

c) Name the reaction mechanism involved in the above test. [2 marks]

d) Hydrogen bromide will react with but-1-ene by this mechanism, producing two isomeric products.

(i) Write a mechanism for the reaction of HBr with $CH_2{=}CHCH_2CH_3$, showing the formation of the major product only. Name the product. [3 marks]

(ii) Explain why it is the major product for this reaction. [2 marks]

This section is free from all GM ingredients...

Wow... these pages are really jam-packed. There's not one, not two, but three mechanisms to learn. And learn 'em you must. And when you've done that, make sure you know both tests for double bonds too. They mightn't be as handy in real life as, say, a tin opener, but you won't need a tin opener in the exam. Unless your exam paper comes in a tin.

E/Z Isomerism

The chemistry on these pages isn't so bad. And don't be too worried when I tell you that a good working knowledge of both German and Latin would be useful. It's not absolutely essential... and you'll be fine without.

Double Bonds Can't Rotate

1) Carbon atoms in a C=C double bond and the atoms bonded to these carbons all lie in the **same plane** (they're **planar**). Because of the way they're arranged, they're actually said to be **trigonal planar** — the atoms attached to each double-bond carbon are at the corners of an imaginary equilateral triangle.

The bond angles in the planar unit are all 120°.

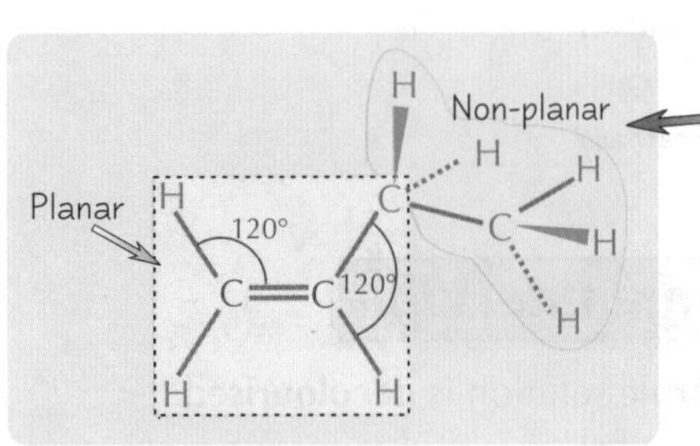

2) Ethene, C_2H_4 (like in the diagram above) is completely planar, but in larger alkenes, only the >C=C< unit is planar.
3) Another important thing about C=C double bonds is that atoms **can't rotate** around them like they can around single bonds (because of the way the p orbitals **overlap** to form a π **bond** — see p44). In fact, double bonds are fairly **rigid** — they don't bend much either.

4) Even though atoms can't rotate about the **double bond**, things can still rotate about any **single bonds** in the molecule — like in this molecule of pent-2-ene.
5) The **restricted rotation** around the C=C double bond is what causes **E/Z isomerism**.

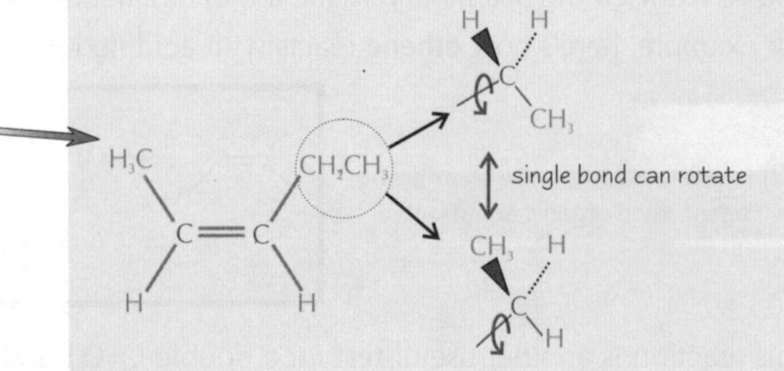

E/Z isomerism is a Type of Stereoisomerism

1) **Stereoisomers** have the same structural formula but a **different arrangement** in space. (Just bear with me for a moment... that will become clearer, I promise.)
2) Because of the **lack of rotation** around the double bond, some **alkenes** can have stereoisomers.
3) Stereoisomers occur when the two double-bonded carbon atoms each have **different atoms** or **groups** attached to them. Then you get an **'E-isomer'** and a **'Z-isomer'**.

   For example, the double-bonded carbon atoms in but-2-ene each have an **H** and a CH_3 group attached.

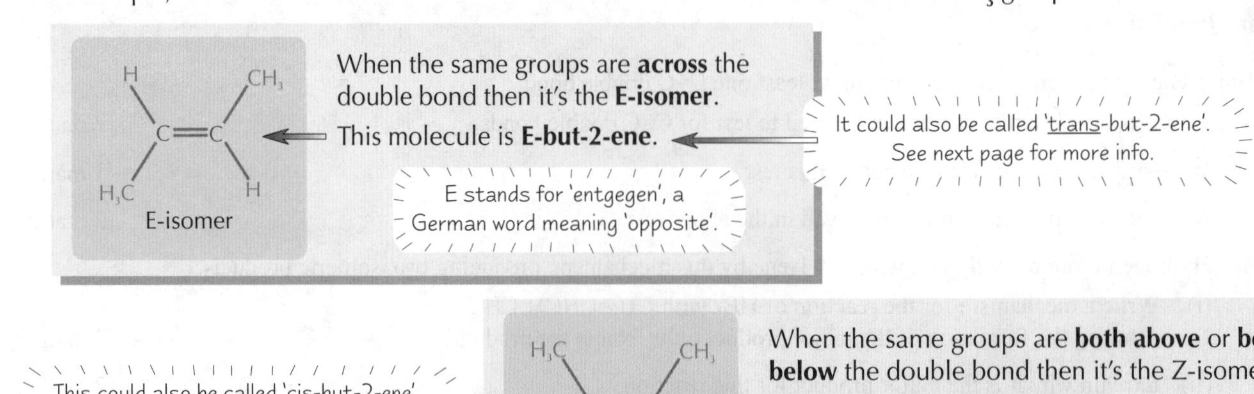

When the same groups are **across** the double bond then it's the **E-isomer**. This molecule is **E-but-2-ene**.

E stands for 'entgegen', a German word meaning 'opposite'.

It could also be called 'trans-but-2-ene'. See next page for more info.

When the same groups are **both above** or **both below** the double bond then it's the Z-isomer. This molecule is **Z-but-2-ene**.

This could also be called 'cis-but-2-ene'. See next page for more info.

Z stands for 'zusammen', the German for 'together'.

E/Z Isomerism

E/Z Isomers Can Sometimes Be Called *Cis-Trans Isomers*

1) E/Z isomerism is sometimes called **cis-trans isomerism**, where...
   (i) '**cis**' means the **Z-isomer**, and
   (ii) '**trans**' means the **E-isomer**.

   So E-but-2-ene can be called trans-but-2-ene, and Z-but-2-ene can be called cis-but-2-ene.

   We're talking Latin this time... 'cis' means 'on the same side', while 'trans' means 'across'.

   Here's another example:
   The **Br** atom and the **CH_3** group are on **opposite** sides of the double bond, so this is **trans-1-bromopropene**.
   No problems there.

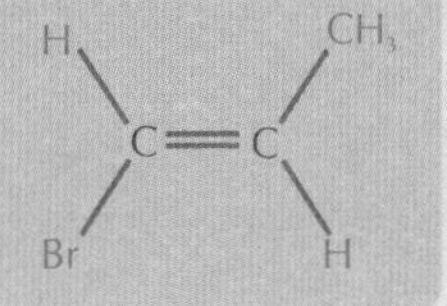

2) But if the carbon atoms both have totally **different** groups attached to them, the cis-trans naming system can't cope.

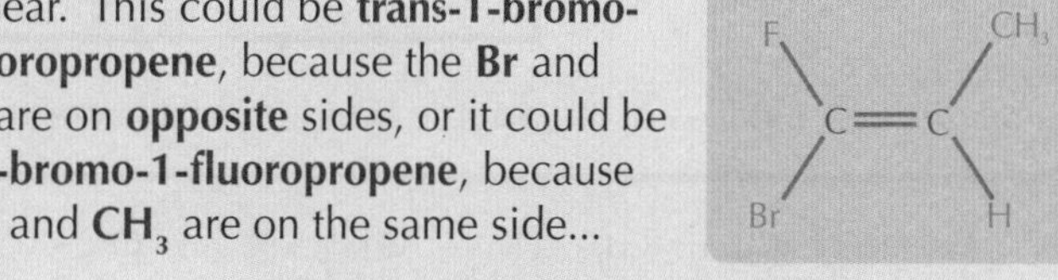

Oh dear. This could be **trans-1-bromo-1-fluoropropene**, because the **Br** and **CH_3** are on **opposite** sides, or it could be **cis-1-bromo-1-fluoropropene**, because the **F** and **CH_3** are on the same side...

3) The E/Z system keeps on working though. This is because each of the groups linked to the double-bonded carbons is given a **priority**.
   If the two carbon atoms have their 'higher priority group' on **opposite** sides, then it's an **E-isomer**.
   If the two carbon atoms have their 'higher priority group' on the **same** side, then it's a **Z-isomer**.

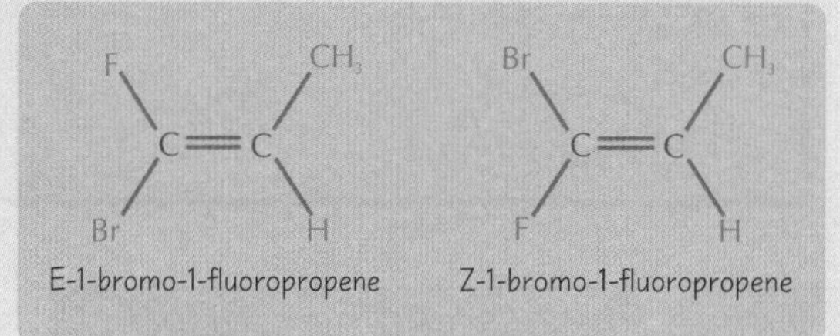

E-1-bromo-1-fluoropropene Z-1-bromo-1-fluoropropene

4) In the E/Z system, Br has a **higher priority** than F, so the names depend on where the Br atom is in relation to the CH_3 group.

You don't need to know the rules for deciding the order of these priorities.

Practice Questions

Q1 Which of the following is the Z-isomer of but-2-ene?

Q2 Define the term 'stereoisomers'.

Q3 Which corresponds to the 'cis-isomer,' the E-isomer or Z-isomer?

Exam Questions

Q1 a) Draw and name the E/Z isomers of pent-2-ene. [4 marks]
b) Explain why alkenes can have E/Z isomers but alkanes cannot. [2 marks]

Q2 An alkene has 4 different groups attached: A, B, X and Y.
Which of the following is the E-isomer if A and X have priority?

1.
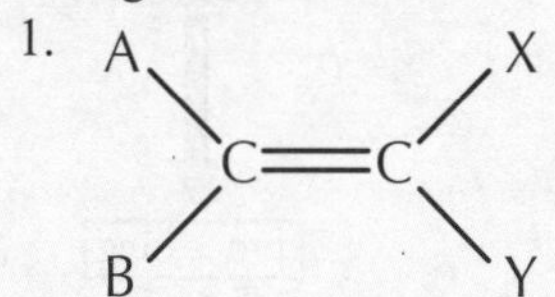

2.
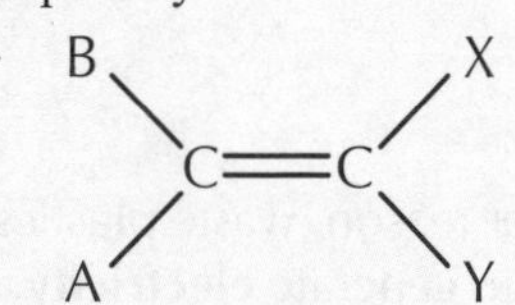

[1 mark]

And there y'all have it, folks — two E/Z pages in an AS Chemistry book...

Cis and trans are fairly easy to remember... 'cis' — think of sisters standing next to each other, while 'trans' means 'across' in things like transmit, transfer, and the Trans-Siberian Railway. And for E/Z isomers, remember that Z-isomers are the ones with the groups on 'ze zame zide'. Or if you prefer, you could learn to speak German...

Polymers

Polymers are long, stringy molecules made by joining lots of little molecules together.

Alkenes **Join Up** to form **Addition Polymers**

1) The **double bonds** in alkenes can open up and join together to make long chains called **polymers**. It's kind of like they're holding hands in a big line. The individual, small alkenes are called **monomers**.
2) This is called **addition polymerisation**. For example, **poly(ethene)** is made by the **addition polymerisation** of **ethene**.

monomer ⟶ polymer

$$n\ \mathrm{H_2C{=}CH_2} \longrightarrow \left(\!-\mathrm{CH_2{-}CH_2}-\!\right)_n$$

The bit in brackets is the 'repeat unit' (or 'repeating unit'). n represents the number of repeat units.

3) Because of the loss of the double bond, poly(alkenes), like alkanes, are **unreactive**.

$$\cdots-\mathrm{CH_2{-}CH(CH_3){-}CH_2{-}CH(CH_3){-}CH_2{-}CH(CH_3)}-\cdots \Rightarrow -\mathrm{CH_2{-}CH(CH_3)}- \Rightarrow \mathrm{H_2C{=}CH(CH_3)}$$

polymer (polypropene) — repeat unit — monomer (propene)

4) To find the **monomer** used to form an addition polymer, take the **repeat unit** and add a **double bond**.

Waste Plastics can be **Buried...**

The **unreactive** nature of most polymers leads to a **problem**. Most polymers aren't **biodegradable**, and so they're really difficult to **dispose of**. Over **2 million** tonnes of plastic waste are produced in the UK each year. It's important to find ways to get rid of this waste while minimising **environmental damage**. There are various possible approaches...

1) You can just **bury** it — take it to a **landfill** site, **compact** it, then cover it with soil. This method's generally used when the plastic is:
   - difficult to separate from other waste,
   - not in sufficient quantities to make separation financially worthwhile,
   - too difficult technically to recycle.
2) But because the **amount of waste** we generate is becoming more and more of a problem, there's a need to **reduce** landfill as much as possible (and there are laws restricting its use).

...or **Recycled...**

Many plastics are made from a finite resource (crude oil) so recycling makes sense. Plastics have to be **sorted** into different types first.

To make sorting easier, plastic products are often marked with numbers for the different polymers, e.g. ♲3 = PVC, and ♲5 = poly(propene)

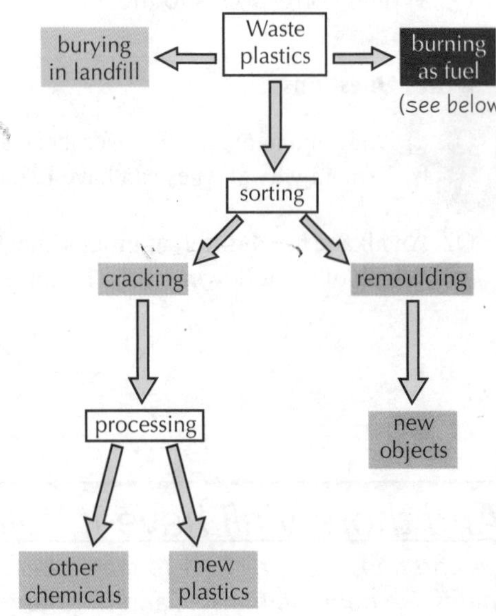

1) Some plastics (poly(propene), for example) can be **melted** and **remoulded**.
2) Some can be **cracked** into **monomers**, and these can be use to make more plastics or other chemicals.

...or **Burned**

1) If recycling isn't possible for whatever reason, waste plastics can be **burned** — and the heat can be used to generate **electricity**.
2) This process needs to be carefully **controlled** to reduce **toxic** gases. For example, polymers that contain **chlorine** (such as **PVC**) produce **HCl** when they're burned — this has to be removed.
3) Waste gases from the combustion are passed through **scrubbers** which can **neutralise** gases such as HCl by allowing them to react with a **base**.

Polymers

Biodegradable Polymers **Decompose** in the **Right Conditions**

Scientists can now make **biodegradable** polymers — ones that naturally **decompose**.

1) **Biodegradable polymers** decompose pretty quickly in certain conditions — because organisms can digest them. (You might get asked about 'compostable' polymers as well as 'biodegradable' ones. These two terms mean more or less the same thing — 'compostable' just means it has to decay fairly quickly, "at the speed of compost".)
2) Biodegradable polymers can be made from materials such as **starch** (from maize and other plants) and from the hydrocarbon **isoprene** (2-methyl-1,3-butadiene). So, biodegradable polymers can be produced from **renewable** raw materials or from **oil fractions**:

> Using **renewable** raw material has several **advantages**.
> (i) Raw materials aren't going to **run out** like oil will.
> (ii) When polymers biodegrade, **carbon dioxide** (a greenhouse gas — see p54) is produced. If your polymer is **plant-based**, then the CO_2 released as it decomposes is the same CO_2 absorbed by the plant when it grew. But with an **oil-based** biodegradable polymer, you're effectively transferring carbon from the oil to the atmosphere.
> (iii) Over their 'lifetime' some plant-based polymers **save energy** compared to oil-based plastics.

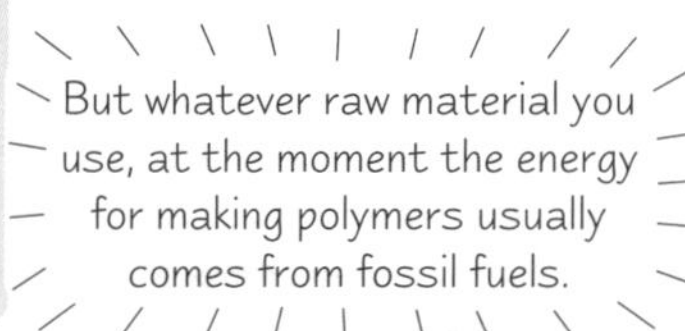

3) Even though they're biodegradable, these polymers still need the right conditions before they'll decompose. You **couldn't** necessarily just put them in a landfill and expect them to perish away — because there's a lack of moisture and oxygen under all that compressed soil. You need to chuck them on a big compost heap. This means that you still need to **collect** and **separate** the biodegradable polymers from non-biodegradable plastics. At the moment, they're also **more expensive** than oil-based equivalents.

4) There are various potential uses — e.g. plastic sheeting used to protect plants from the frost can be made from poly(ethene) with **starch grains** embedded in it. In time the starch is broken down by **microorganisms** and the remaining poly(ethene) crumbles into dust. There's no need to collect and dispose of the old sheeting.

Practice Questions

Q1 Draw the displayed formulas for the monomer and repeat unit used to make poly(propene).

Q2 Name a renewable material that a polymer can be made from.

Q3 Describe three ways in which used polymers such as poly(propene) can be disposed of.

Exam Questions

Q1 Part of the structure of a polymer is shown on the right.

a) Draw the repeating unit of the polymer. [1 mark]

b) Draw the monomer from which the polymer was formed. [1 mark]

```
 H   H    H   H    H
 |   |    |   |    |
-C---C----C---C----C-
 |   |    |   |    |
 H  C6H5  H  C6H5  H
```

Q2 Waste plastics can be disposed of by burning.

a) Describe one advantage of disposing of waste plastics by burning. [1 mark]

b) Describe a disadvantage of burning waste plastic that contains chlorine, and explain how the impact of this disadvantage could be reduced. [2 marks]

Q3 Apart from being biodegradable, describe TWO benefits of using starch- or maize-based polymers instead of oil-based polymers. [2 marks]

Phil's my plastic plane — but I don't have permission to land Phil...

You might have noticed that recycling and biodegradability are hot topics these days. And not just in the usual places, such as Chemistry books. No, this kind of stuff regularly makes it onto the news. It's also a good tip for small talk at parties. (I'm not joking — at a party last week, I met someone who researches biodegradable plastics for a living. He told me all about it.)

Shapes of Molecules

Chemistry would be heaps more simple if all molecules were flat. But sadly they're not.

Molecular Shape depends on Electron Pairs Around the Central Atom

Molecules and molecular ions come in loads of **different shapes**.

The shape depends on the **number of pairs** of electrons in the outer shell of the central atom.

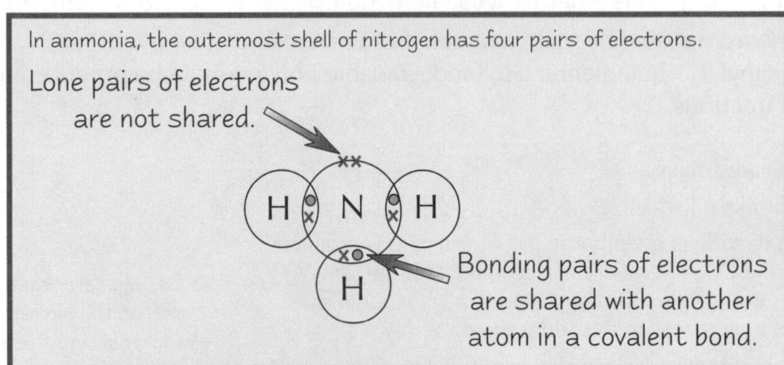

A lone pear

Electron Pairs Repel Each Other

1) Electrons are all **negatively charged**, so it's pretty obvious that electron pairs will **repel** each other as much as they can.
2) This sounds straightforward, but the **type** of the electron pair affects **how much** it repels other electron pairs. Lone pairs repel **more** than bonding pairs.
3) So, the **greatest** angles are between **lone pairs** of electrons, and bond angles between bonding pairs are often **reduced** because they are pushed together by lone-pair repulsion.

Lone-pair/lone-pair bond angles are the biggest.	*Lone-pair/bonding-pair bond angles are the second biggest.*	*Bonding-pair/bonding-pair bond angles are the smallest.*

4) This is known by the long-winded name '**electron-pair repulsion theory**'.

The central atoms in these molecules all have **four pairs** of electrons in their outer shells, but they're all **different shapes.**

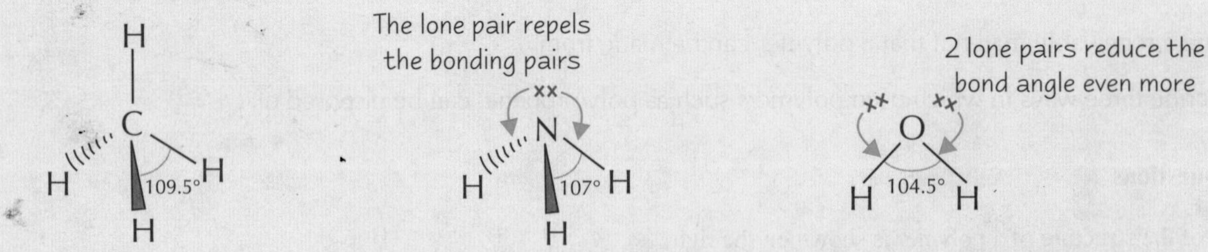

The lone pair repels the bonding pairs

2 lone pairs reduce the bond angle even more

109.5°

107°

104.5°

Methane — no lone pairs

Ammonia — 1 lone pair

Water — 2 lone pairs

Single-Bonded Carbon Atoms Have Their Bonds Arranged Like a Tetrahedron

1) When a carbon atom makes four single bonds (as in alkanes), the atoms around each carbon form a **tetrahedral shape**.
2) The angle between any two of the covalent bonds is **109.5°**. At this angle, the bonds (i.e. the electron pairs) are as **far apart** from each other as possible.

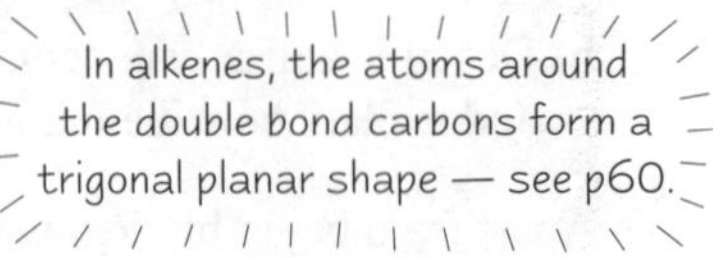

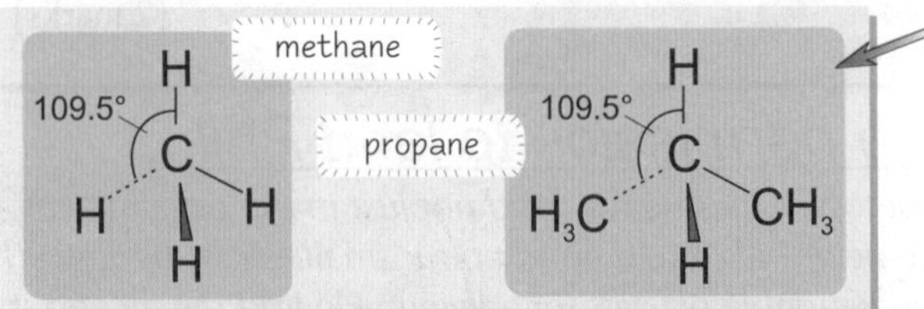

3) This **tetrahedral** shape around each carbon atom means that single-bonded carbon chains containing 3 or more carbon atoms form a 'wiggly line'.

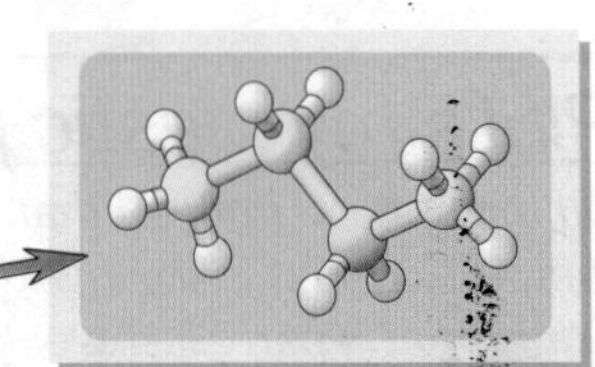

Shapes of Molecules

Practise Drawing these Molecules

You need to be able to **explain the shapes** of all these molecules (and any that are basically the same, like BF_3 instead of BCl_3).

2 ELECTRON PAIRS ON CENTRAL ATOM —

$BeCl_2$ Cl—Be—Cl 180°

CO_2 O=C=O 180°

Linear molecules

Just treat double bonds the same as single bonds (even though there might be slightly more repulsion from a double bond).

3 ELECTRON PAIRS ON CENTRAL ATOM —

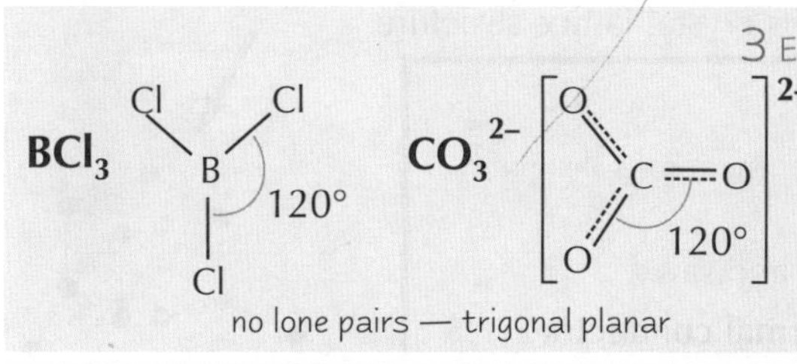

no lone pairs — trigonal planar

(in CO_3^{2-} and NO_3^- the bonds are all midway between single and double bonds)

NO_3^- O, N=O, O, 120°

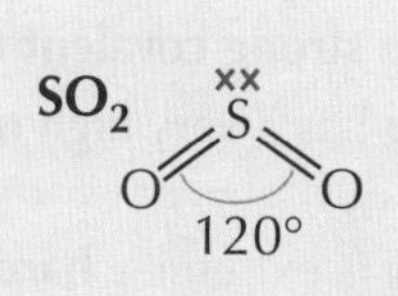

Here, the extra electron density in the double bonds cancels out the extra repulsion from the lone pair, so you still get 120°.

1 lone pair — non-linear or 'bent'

4 ELECTRON PAIRS ON CENTRAL ATOM —

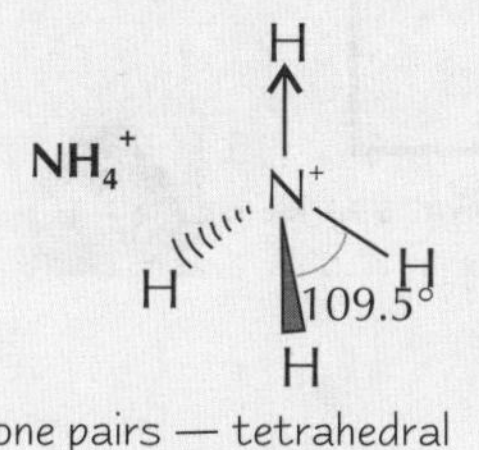

no lone pairs — tetrahedral

NH_3 N, H, H, H, 107°

SO_3^{2-} S, O⁻, O⁻, O, 107°

1 lone pair — trigonal pyramidal

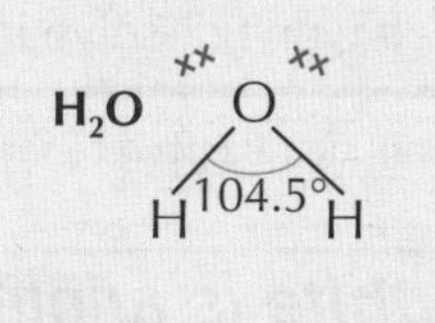

2 lone pairs — non-linear or 'bent'

Some central atoms can use d orbitals and can 'expand the octet' — which means they can have more than eight bonding electrons. E.g. in PCl_5, phosphorus has 10 electrons in its outermost shell, while in SF_6, sulfur has 12.

5 ELECTRON PAIRS ON CENTRAL ATOM —

PCl_5 P, Cl, Cl, Cl, Cl, Cl, 90°, 120°

no lone pairs — trigonal bipyramidal

6 ELECTRON PAIRS ON CENTRAL ATOM —

SF_6 S, F, F, F, F, F, F, All bond angles 90°

no lone pairs — octahedral

Practice Questions

Q1 What is a lone pair of electrons?

Q2 Write down the order of the strength of repulsion between different kinds of electron pairs.

Q3 Describe the difference in shape between an alkane and an alkene.

Q4 What shape is PCl_5?

Q5 Draw a tetrahedral molecule.

Exam Question

Q1 Nitrogen and boron can form the chlorides NCl_3 and BCl_3.

a) Draw 'dot and cross' diagrams to show the bonding in NCl_3 and BCl_3. [2 marks]

b) Draw the shapes of the molecules NCl_3 and BCl_3.
Show the approximate values of the bond angles on the diagrams and name each shape. [6 marks]

c) Explain why the shapes of NCl_3 and BCl_3 are different. [3 marks]

These molecules ain't square...

In the exam, those evil examiners might try to throw you by asking you to predict the shape of an unfamiliar molecule. Don't panic — it'll be just like one you do know, e.g. PH_3 is the same shape as NH_3. Make sure you can draw every single molecule on this page. Yep, that's right — from memory. And you need to know what the shapes are called too.

Carbon Structures

Diamonds, besides being a girl's best friend, are made of carbon atoms arranged in a giant molecular structure. But diamond isn't the only giant molecular structure made of carbon...

Diamond is the Hardest known Substance

Allotropes are different forms of the **same element** in the **same state**. Carbon forms **three** allotropes — **diamond**, **graphite** and **fullerenes**. Each allotrope has a **different** giant molecular structure.

Diamond is made up of **carbon atoms**. Each carbon atom is **covalently bonded** with σ bonds to **four** other carbon atoms. The atoms arrange themselves in a **tetrahedral** shape — its crystal lattice structure.

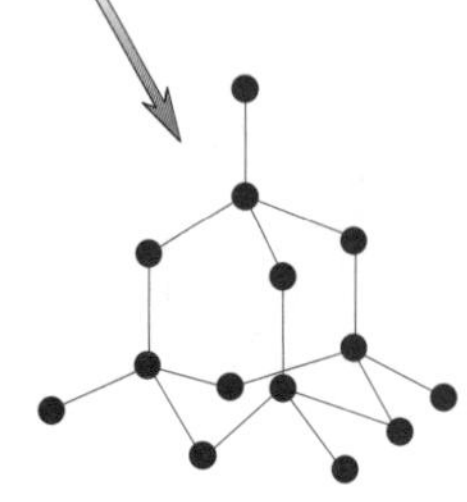

Because of its **strong covalent** bonds:

1) Diamond has a **very high melting point** — it actually sublimes (changes straight from a solid to a gas, skipping out the liquid stage) at over 3800 K.
2) Diamond is extremely **hard** — it's used in diamond-tipped drills and saws.
3) **Vibrations** travel easily through the stiff lattice, so it's a **good thermal conductor**.
4) It **can't conduct** electricity — all the outer electrons are held in localised bonds.
5) It won't dissolve in **any** solvent.

You can 'cut' diamond to form gemstones. The regular structure makes it refract light a lot, which is why it sparkles.

Graphite is another Allotrope of Carbon

Graphite has a **different macromolecular structure** from diamond:

1) The carbon atoms are arranged in **sheets** of flat hexagons covalently bonded with three bonds each.
2) The fourth outer electron of each carbon atom is **delocalised**.

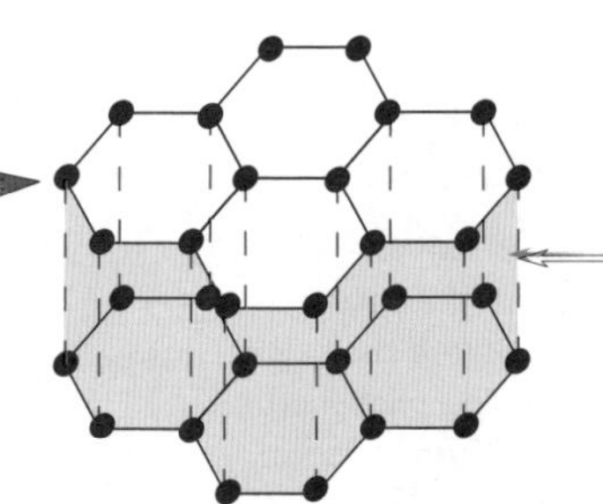

The sheets of hexagons are bonded together by weak intermolecular bonds (London forces).

Graphite's structure means that it has some **different properties** from diamond (and a few similar ones too):

1) The weak bonds **between** the layers in graphite are easily broken, so the sheets can slide over each other — graphite feels **slippery** and is used as a **dry lubricant** and in **pencils**.
2) The '**delocalised**' electrons in graphite aren't attached to any particular carbon atoms and are **free to move** along the sheets, so an **electric current** can flow.
3) The layers are quite **far apart** compared to the length of the covalent bonds, so graphite is **less dense** than diamond and is used to make **strong, lightweight** sports equipment.
4) Because of the **strong covalent bonds** in the hexagon sheets, graphite also has a **very high melting point** (it sublimes at over 3900 K).
5) Like diamond, graphite is **insoluble** in any solvent. The covalent bonds in the sheets are **too difficult** to break.

The Fullerenes Include Hollow Balls...

1) **Fullerenes** are **molecules** of carbon shaped like **hollow balls** or **tubes**. Each carbon atom forms **three** covalent bonds with its neighbours, leaving **free electrons** that can **conduct** electricity.
2) Fullerenes are **nanoparticles**. Nanoparticles are generally up to 100 nanometres across.
3) The first fullerene to be discovered was **buckminsterfullerene**, which has **60** carbon atoms joined to make a **ball** — its molecular formula is C_{60}. It occurs naturally in soot.
4) Many fullerenes are **soluble** in **organic solvents**, and form brightly coloured solutions.
5) Because they're hollow, fullerenes can be used to '**cage**' other molecules. The fullerene structure forms around another molecule, which is then trapped inside. This could be used as a way of **delivering a drug** into specific cells of the body.
6) Fullerenes are used in **nanotechnology** — materials and devices made from **nanoparticles**. At this tiny scale, materials often have very **different properties** from '**bulk**' forms of the same substance.

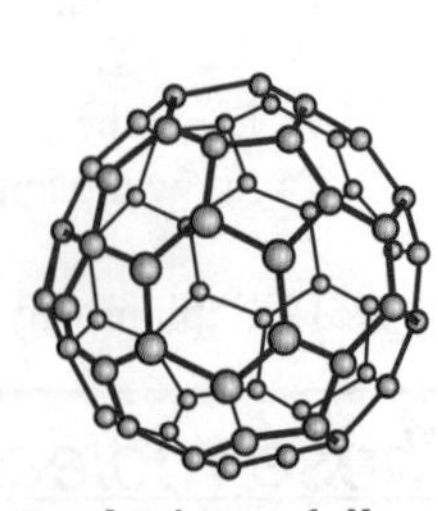

Buckminsterfullerene

Carbon Structures

...And Tubes

A **carbon nanotube** is like a single layer of graphite rolled up into a tiny hollow cylinder.

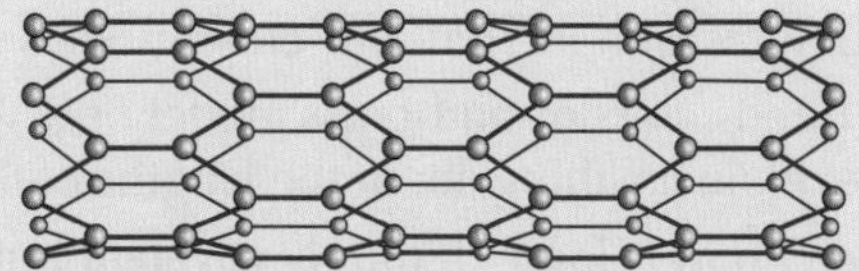

1) All those covalent bonds make carbon nanotubes **very strong**. They can be used to reinforce graphite in tennis rackets and to make stronger, lighter building materials.
2) Nanotubes **conduct** electricity, so they can be used as tiny wires in **circuits** for computer chips.
3) The ends of a nanotube can be **'capped'**, or closed off, to create a large molecular cage structure.

There's a Debate About the Safety of Nanotechnology

When someone comes up with a **new** way to use **nanotechnology** there are often questions about whether it's **safe**. As with all practical applications of science it's a question of weighing up any **risks** against the possible **benefits**:

1) A new nanotechnology could improve people's **health** or **quality of life**.
2) Before they're used, new products and technologies are **thoroughly tested** to make sure they're **not harmful**.
3) Some people question whether any nanotechnology has been tested enough to **prove** that it's **safe**. Others think that where the **benefits** are big enough, we should take the **risk** of starting to use nanotechnologies that **seem** to be safe.

Example — Nanoparticles in Transparent Sunscreen Creams

1) Many **sunscreen** creams contain nanoparticles of **zinc oxide** and **titanium dioxide**. These particles **reflect UV radiation**, but at this size they appear **transparent** — so the sunscreen can be **see-through.**
2) People are much more likely to use sunscreen if it's see-through. And it's really important that they do, because it reduces their risk of getting **skin cancer**.
3) Scientists have found that some nanoparticles seem to be able to **pass through** the **skin** and into **cells**. And if they can do that, there's a possibility that they could **damage** the cells.
4) Researchers studying the sunscreens **don't think** that these nanoparticles are capable of getting through the skin. But some people think we should **avoid using them** until we're **absolutely sure** that they're not harmful.

Practice Questions

Q1 What are fullerenes?

Q2 Describe the structure of nanotubes.

Exam Questions

Q1 a) Explain, in terms of bonding, the following observations:

(i) Graphite is a conductor of electricity but diamond is not. [4 marks]

(ii) Graphite is less dense than diamond. [2 marks]

(iii) Graphite is slippery and can be used as a lubricant but diamond is the hardest substance known. [3 marks]

b) Give a physical property that is shared by both diamond and graphite. [1 mark]

Q2 a) Describe the structure of buckminsterfullerene. [3 marks]

b) Carbon nanotubes are a member of the fullerene family. Explain one potential use of carbon nanotubes. [2 marks]

Y'all say howdy to buckminsterfullerene — the state molecule of Texas...

No really, it is. Not that you're going to get any marks for knowing that. But you'll probably get some for knowing about its structure and uses. That goes for carbon nanotubes too. And you should definitely know the properties of diamond and graphite and how they are explained by their structures. Because they're things that examiners love asking questions about.*

* Unless you're appearing in a seriously obscure quiz show.

Electronegativity and Polarisation

Opposites attract, like Jack Sprat and his wife — that's all you need to know. Well OK, that's not true, so get learnin'...

There are **Limitations** to **Models** of **Bonding**

There are some problems with the **dot-and-cross** models of ionic and covalent bonding.

1) With covalent bonds, the **dot-and-cross** model only illustrates how the atoms in a compound **share** their electron pairs. It can't explain anything about the **lengths** of the covalent bonds formed, or the overall **shape** of a molecule.
2) Most bonds aren't **purely ionic** or **purely covalent** but somewhere in between. Thanks to **bond polarisation** (see page 43 and below) most compounds have a **mixture** of ionic and covalent properties.

There's a Gradual **Transition** from Ionic to Covalent Bonding

1) Very few compounds come even close to being **purely ionic**.
2) Only bonds between atoms of a **single** element, like diatomic gases such as hydrogen (H_2) or oxygen (O_2), can be **purely covalent**.
3) 'Purely ionic' and purely covalent bonding are the **extremes** of a **continuum** (a continuous scale). Most compounds come somewhere **in between** these two extremes — meaning they've often got ionic **and** covalent properties, e.g. covalent hydrogen chloride gas molecules dissolve to form hydrochloric acid, which is an ionic solution.

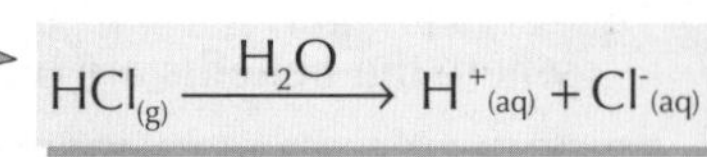

$$HCl_{(g)} \xrightarrow{H_2O} H^+_{(aq)} + Cl^-_{(aq)}$$

Some Atoms **Attract** Bonding Electrons More than Other Atoms

The ability to attract the bonding electrons in a covalent bond is called electronegativity.

1) Electronegativity is usually measured using the **Pauling scale**.
2) **Fluorine** is the most electronegative element — it's given a value of **4.0** on the Pauling scale. Oxygen, nitrogen and chlorine are also very strongly electronegative.

Element	H	C	N	Cl	O	F
Electronegativity	2.1	2.5	3.0	3.0	3.5	4.0

Covalent Bonds may be Polarised by **Differences** in **Electronegativity**

In a covalent bond between two atoms of **different** electronegativities, the bonding electrons are **pulled towards** the more electronegative atom. This makes the bond **polar**.

1) The covalent bonds in diatomic gases (e.g. H_2, Cl_2) are **non-polar** because the atoms have **equal** electronegativities and so the electrons are equally attracted to both nuclei.

H—°ₓ—H

Permanent polar bonding

2) Some elements, like carbon and hydrogen, have pretty **similar** electronegativities, so bonds between them are essentially **non-polar**.
3) In a **polar bond**, the difference in electronegativity between the two atoms causes a **dipole**. A dipole is a **difference in charge** between the two atoms caused by a shift in **electron density** in the bond.

$-\overset{|}{\underset{|}{C}}^{\delta+}$—°ₓ—$Br^{\delta-}$

'δ' (delta) means 'slightly', so 'δ+' means 'slightly positive'.

4) If the difference is large enough, the bond becomes **pretty much ionic**.
5) So what you need to **remember** is that the greater the **difference** in electronegativity, the **more polar** the bond.

Electronegativity and Polarisation

Polar Bonds **Don't** always make **Polar Molecules**

Some molecules with polar bonds are **polar molecules** — the molecule itself has a **permanent dipole**. Whether a molecule itself is polar depends on its **shape** and the **polarity** of its bonds.

1) In a simple molecule, such as **hydrogen chloride**, the polar bond gives the whole molecule a permanent dipole — it's a **polar molecule**.

$\overset{\delta+}{H}-\overset{\delta-}{Cl}$ polar

2) A more complicated molecule may have **several polar bonds**. If the polar bonds are arranged so they point in opposite directions, they'll **cancel each other out** — the molecule is **non-polar** overall.

No dipole overall. $\overset{\delta-}{O}=\overset{\delta+}{C}=\overset{\delta-}{O}$

3) If the polar bonds all point in roughly the **same direction**, then the molecule will be **polar**.

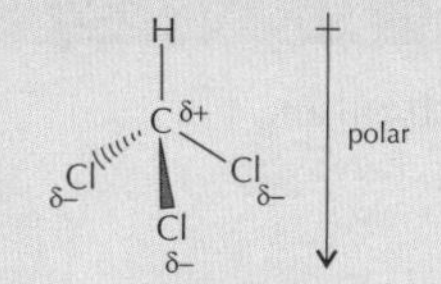

4) **Lone pairs of electrons** on the central atom also have an effect on the overall polarity and may **cancel out** the dipole created by the bonding pairs.

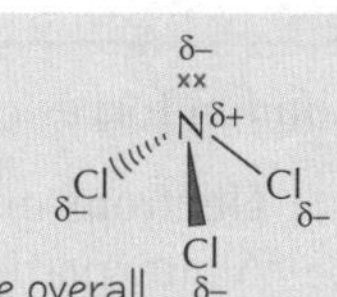

The **Length** of a Bond is Related to its **Strength**

1) In covalent bonds, there isn't just an **attraction** between the nuclei and the shared electrons. The two **positively charged nuclei** also **repel** each other, as do the **electrons**.
2) The distance between the **two nuclei** is the distance where the **attractive** and **repulsive** forces balance each other. This distance is the **bond length**.
3) The **stronger** the attraction between the atoms, the higher the **bond enthalpy** and the **shorter** the bond length. It makes sense really. If there's more attraction, the nuclei will pull **closer** together.

A C=C bond has a greater bond enthalpy and is shorter than a C–C bond. Four electrons are shared in C=C and only two in C–C, so the electron density between the two carbon atoms is greater. C≡C has an even higher bond enthalphy and is shorter than C=C — six electrons are shared here.

Bond	C–C	C=C	C≡C
Average Bond Enthalpy (kJ mol^{-1})	+347	+612	+838
Bond length (nm)	0.154	0.134	0.120

Practice Questions

Q1 What are the only bonds which can be purely covalent?

Q2 What is the most electronegative element?

Q3 Explain why the molecule H-Br has a permanent dipole.

Q4 Explain why the molecule dichloromethane (CH_2Cl_2) has a dipole.

Exam Question

Q1 Many covalent molecules have a permanent dipole, due to differences in electronegativities.

a) Define the term electronegativity. [2 marks]

b) Draw the shapes and predict the overall polarity of the following molecules, marking any bond polarities clearly on your diagram:
(i) Br_2 (ii) H_2O (iii) CCl_4 (iv) NH_3 [8 marks]

c) Fluorine is the most electronegative element.
NF_3 is the same shape as NH_3, yet it has no permanent dipole. Why is this? [2 marks]

Models have limitations — well, we can't all be pretty **and** clever...

Understanding different types of bond is pretty important. When you were doing GCSEs, people probably told you that bonds were either ionic, covalent or metallic — nice and simple. Sadly, they lied. There's a sliding scale of ionic-ness and covalent-ness, and it's all to do with electronegativity and polarisation. You might have to explain it — make sure you can.

Intermolecular Forces

Intermolecular forces hold molecules together. They're pretty important, cos we'd all be gassy clouds without them.

Intermolecular Forces are **Very Weak**

Intermolecular forces are forces **between** molecules. They're much **weaker** than covalent, ionic or metallic bonds. There are three types you need to know about:

1) **Instantaneous dipole-induced dipole** or **London** forces (this is the weakest type)
2) **Permanent dipole-dipole interactions**
3) **Hydrogen bonding** (this is the strongest type)

London forces are also known as van der Waals forces. (And sometimes the term 'van der Waals forces' is considered to include all three types of intermolecular force.)

London Forces are Found Between **All** Atoms and Molecules

London forces cause **all** atoms and molecules to be **attracted** to each other.

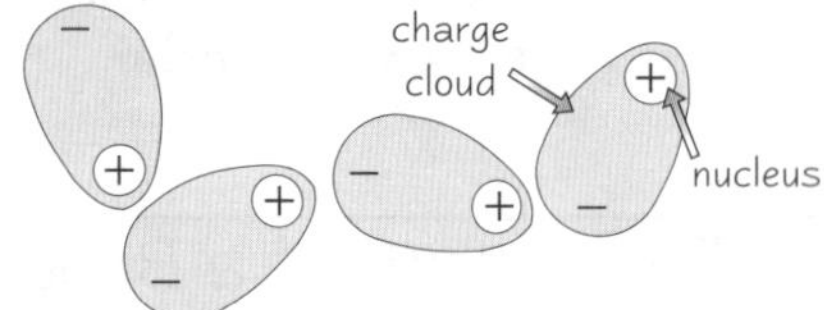

1) **Electrons** in charge clouds are always **moving** really quickly. At any particular moment, the electrons in an atom are likely to be more to one side than the other. At this moment, the atom would have a **temporary dipole**.
2) This dipole can cause **another** temporary dipole in the opposite direction on a neighbouring atom. The two dipoles are then **attracted** to each other.
3) The second dipole can cause yet another dipole in a **third atom**. It's kind of like a domino rally.
4) Because the electrons are constantly moving, the dipoles are being **created** and **destroyed** all the time. Even though the dipoles keep changing, the **overall effect** is for the atoms to be **attracted** to each another.

Stronger **London Forces** mean **Higher Melting and Boiling Points**

1) Not all London forces are the same strength — larger molecules have **larger electron clouds**, meaning **stronger** London forces.
2) Molecules with greater **surface areas** also have stronger London forces because they have a **bigger exposed electron cloud**.
3) When you **boil** a liquid, you need to **overcome** the intermolecular forces, so that the particles can **escape** from the liquid surface. It stands to reason that you need **more energy** to overcome **stronger** intermolecular forces, so liquids with stronger London forces will have **higher boiling points**.
4) Melting solids also involves **overcoming intermolecular forces**, so solids with stronger London forces will have **higher melting points** too.
5) Alkanes demonstrate this nicely...

Boiling Points of Alkanes Depend on **Size** and **Shape**

The smallest alkanes, like methane, are **gases** at room temperature and pressure — they've got very low boiling points. Larger alkanes are **liquids** — they have higher boiling points.

1) Alkanes have **covalent bonds** inside the molecules. **Between** the molecules, there are **London forces** which hold them all together.
2) The **longer** the carbon chain, the **stronger** the London forces — there's **more molecular surface area** and more electrons to interact.
3) Branched-chain alkanes have smaller **molecular surface areas** and they can't **pack as closely** together — so the London forces are reduced.
4) So:

- **Long-chain alkanes have higher boiling points than short-chain alkanes.**
- **Straight-chain alkanes have higher boiling points than branched alkanes.**

5) The **melting temperatures** of alkanes vary in a similar way, for the same reasons.

Example: Isomers of C_4H_{10}

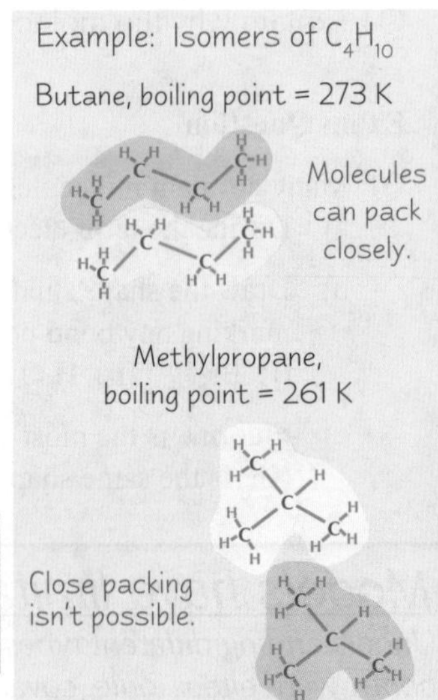

Intermolecular Forces

Polar Molecules have Permanent Dipole-Dipole Forces

The $\delta+$ and $\delta-$ charges on **polar molecules** cause **weak electrostatic forces** of attraction **between** molecules.

E.g. hydrogen chloride gas has polar molecules.

$\overset{\delta+}{H}-\overset{\delta-}{Cl}\cdots\overset{\delta+}{H}-\overset{\delta-}{Cl}\cdots\overset{\delta+}{H}-\overset{\delta-}{Cl}$

Even though they're weak, the forces are still much stronger than London forces.

Now this bit's pretty cool:

If you put an **electrostatically charged rod** next to a jet of a polar liquid, like water, the liquid will **move** towards the rod. I wouldn't believe me either, but it's true. It's because **polar liquids** contain molecules with **permanent dipoles**. It doesn't matter if the rod is **positively** or **negatively** charged. The polar molecules in the liquid can **turn around** so the oppositely charged end is attracted towards the rod.

You can use this experiment to find out if the molecules of a jet of liquid are **polar or non-polar**.

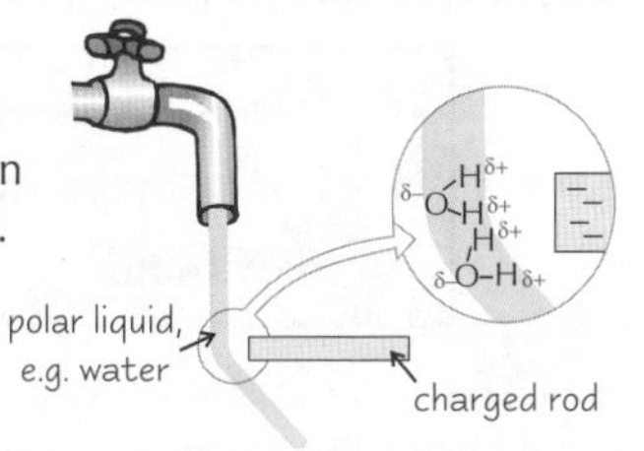

Hydrogen Bonding is the Strongest Intermolecular Force

1) Hydrogen bonding **only** happens when **hydrogen** is covalently bonded to **fluorine**, **nitrogen** or **oxygen**.
2) Fluorine, nitrogen and oxygen are very **electronegative**, so they draw the bonding electrons away from the hydrogen atom. The bond is so **polarised**, and hydrogen has such a **high charge density** because it's so small, that the hydrogen atoms form weak bonds with **lone pairs of electrons** on the fluorine, nitrogen or oxygen atoms of **other molecules**.
3) Molecules which have hydrogen bonding are usually **organic**, containing **-OH** or **-NH** groups. **Water** and **ammonia** both have hydrogen bonding.

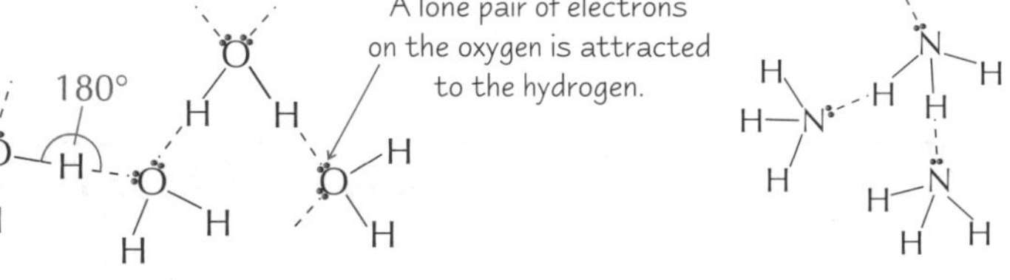

4) Hydrogen bonding has a **huge effect** on the properties of substances.
   - They have **higher boiling and melting points** than other similar molecules because of the **extra energy** needed to break the hydrogen bonds.

     This is the case with **water**, and also **hydrogen fluoride**, which has a much **higher boiling point** than the other hydrogen halides.
   - Ice has more hydrogen bonds than liquid water, and hydrogen bonds are relatively **long**. So the H_2O molecules in ice are further apart on average, making ice **less dense** than liquid water.

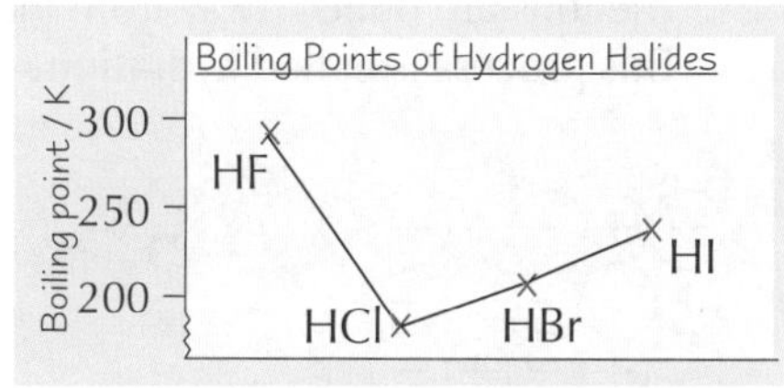

Practice Questions

Q1 What's the strongest type of intermolecular force?

Q2 What is a hydrogen bond?

Exam Question

Q1 Water, H_2O, boils at 373 K.

a) Draw a clearly labelled diagram to show all the forms of intra- and intermolecular bonding in water. [4 marks]

Intra-molecular bonding is bonding inside molecules.

b) Group 6 hydrides have the general formula H_2X. The graph shows the boiling points of Group 6 hydrides. Explain the trends shown in terms of intermolecular forces. [4 marks]

Boiling Points of Group 6 Hydrides
Boiling point / K: 400, 300, 250, 200
H_2O, H_2S, H_2Se, H_2Te

Q2 a) Draw dot-and-cross diagrams for the following molecules: (i) F_2, (ii) HF [2 marks]

b) Explain why only one of these molecules has a permanent dipole. [2 marks]

c) One of the substances in a) is a liquid at room temperature and the other is a gas. State which one is a liquid and explain your reasoning. [2 marks]

London Forces — the irresistible pull of streets paved with gold...

Just because intermolecular forces are a bit wimpy and weak, don't forget they're there. It'd all fall apart without them. Learn the three types — London forces, permanent dipole-dipole forces and hydrogen bonds. I bet fish are glad that water forms hydrogen bonds. If it didn't, their water would boil. (And they wouldn't have evolved in the first place.)

Solubility

Ooh, solubility. Alcohol dissolves in water, which is fortunate for all those Scotch whisky connoisseurs who like their single malt with 'a wee drop of water in it'. But this page is so much more satisfying than whisky and water.

Solubility is Affected by Bonding

1) For one substance to **dissolve** in another, all these things have to happen:

- bonds in the **substance** have to **break**,
- bonds in the **solvent** have to **break**, and
- **new bonds** have to form **between** the **substance** and the **solvent**.

2) Usually a substance will only dissolve if the strength of the new bonds **formed** is about **the same as**, or **greater than**, the strength of the bonds that are **broken**.

There Are Polar and Non-polar Solvents

There are two main **types of solvent**:

1) **Polar solvents** such as water. Water molecules bond to each other with **hydrogen bonds**.
2) **Non-polar solvents** such as hexane. Hexane molecules bond to each other by **London forces**.

Many substances are soluble in one type of solvent but not the other — and you'll be expected to understand why...

Ionic Substances Dissolve in Polar Solvents such as Water

1) The ions are attracted to the **oppositely charged ends** of the water molecules.
2) The ions are pulled away from the ionic lattice by the water molecules, which surround the ions. This process is called **hydration**.

ions in a lattice

polar water molecules

hydrated ions

3) Some ionic substances **don't dissolve** because the bonding between their ions is **too strong**. E.g. Aluminium oxide (Al_2O_3) is insoluble in water because the bonds between the ions are stronger than the bonds they'd form with the water molecules. (Al^{3+} has a high charge density, so it's highly polarising — see p43.)

Alcohols also Dissolve in Polar Solvents such as Water

1) Alcohols are **covalent** but they dissolve in water...
2) ... because the polar O-H bond in an alcohol is attracted to the polar O-H bonds in water. **Hydrogen bonds** form between the lone pairs on the δ- oxygen atoms and the δ+ hydrogen atoms.

3) The **carbon chain** part of the alcohol isn't attracted to water, so the more carbon atoms there are, the **less soluble** the alcohol will be.

Henry couldn't understand why the champagne wouldn't dissolve.

Solubility

Not All **Molecules** with **Polar Bonds** Dissolve in **Water**

1) **Halogenoalkanes** contain **polar bonds** but their dipoles aren't strong enough to form **hydrogen bonds** with water. (See p71 for what's needed to form hydrogen bonds.)
2) The hydrogen bonding **between** water molecules is **stronger** than the bonds that would be formed with halogenoalkanes, so halogenoalkanes don't dissolve.

Example:
When the halogenoalkane chlorobutane is added to water, they don't mix, but separate into two layers.

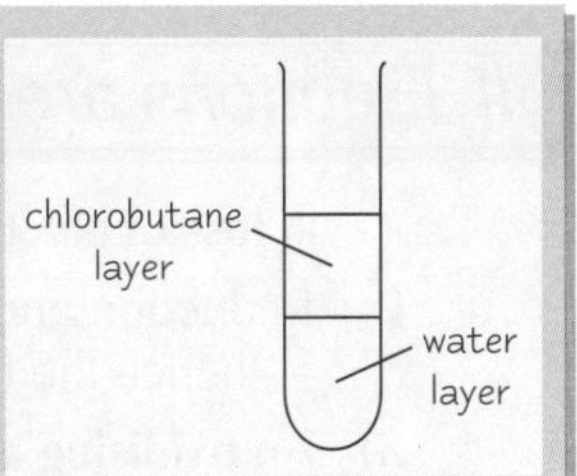

Non-polar Substances Dissolve Best in **Non-polar Solvents**

1) Non-polar substances such as **ethene** have **London forces** between their molecules. They form **similar bonds** with **non-polar solvents** such as hexane — so they tend to dissolve in them.
2) Molecules of **polar solvents** such as water are attracted to **each other** more strongly than they are to **non-polar molecules** such as iodine — so non-polar substances don't tend to dissolve easily in polar solvents.

Like dissolves like (usually) — substances usually dissolve best in solvents that have **similar bonds.**

Practice Questions

Q1 Which type of solvent, polar or non-polar, would you choose to dissolve

i) sodium chloride? ii) ethane?

Q2 Why do most ionic substances dissolve in water?

Q3 Some ionic substances don't dissolve in water. Why not?

Q4 What is meant by 'hydration'?

Q5 What type of bonding occurs between an alcohol and water?

Q6 Why are most non-polar substances insoluble in polar solvents?

Exam Questions

Q1 Hydrogen bonds are present between molecules of water.

a) (i) Explain why alcohols often dissolve in water while halogenoalkanes do not. [4 marks]

(ii) Draw a labelled diagram to show the bonds that form when propanol dissolves in water. [2 marks]

b) Explain the process by which potassium iodide dissolves in water to form hydrated ions. Include a diagram of the hydrated ions. [5 marks]

Q2 a) Describe how you could determine that an unknown substance, X, was likely to be a non-polar covalent compound by testing with two different solvents. Name the solvents chosen and give the expected results. [4 marks]

b) Explain these results in terms of the bonding within X and the solvents. [4 marks]

Daddy, when the ice-caps melt, where will all the polar solvents live?

I reckon it's logical enough, this business of what dissolves what. Remember, water is a polar molecule — so other polar molecules, as well as ions, are attracted to its δ^+ and δ^- ends. If that attraction's stronger than the existing bonds (which have to break), the substance will dissolve. It's worth remembering that rule of thumb about 'like dissolves like'.

Oxidation and Reduction

This next bit has more occurences of "oxidation" than the Beatles' "All You Need is Love" features the word "love".

If Electrons are Transferred, it's a **Redox Reaction**

1) A **loss** of electrons is called **oxidation**. A **gain** in electrons is called **reduction**.
2) Reduction and oxidation happen **simultaneously** — hence the term "**redox**" reaction.
3) An **oxidising agent accepts** electrons and gets reduced.
4) A **reducing agent donates** electrons and gets oxidised.

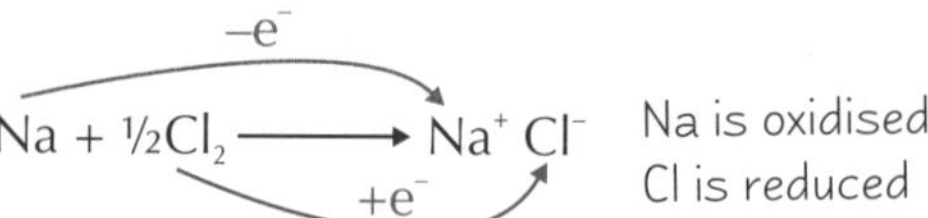

Sometimes it's easier to talk about **Oxidation Numbers**

(They're also called oxidation states.)

There are lots of rules. Take a deep breath...

1) All atoms are treated as **ions** for this, even if they're covalently bonded.

2) Uncombined **elements** have an oxidation number of **0**.

3) Elements just bonded to **identical atoms**, like O_2 and H_2, also have an oxidation number of **0**.

4) The oxidation number of a simple **monatomic ion**, e.g. Na^+, is the same as its **charge**.

5) In **compounds** or **compound ions**, the **overall oxidation number** is just the ion charge.

SO_4^{2-} — **overall oxidation number = -2**,
oxidation number of **O = -2** (total = -8),
so oxidation number of **S = +6**

Within an ion, the most electronegative element has a negative oxidation number (equal to its ionic charge). Other elements have more positive oxidation numbers.

6) The sum of the oxidation numbers for a **neutral compound** is 0.

Fe_2O_3 — **overall oxidation number = 0**, oxidation number of **O = -2** (total = -6), so oxidation number of **Fe = +3**

7) Combined oxygen is nearly always -2, except in peroxides, where it's -1, (and in the fluorides OF_2, where it's +2, and O_2F_2, where it's +1 (and O_2 where it's 0).)

In H_2O, oxidation number of **O = -2**, but in H_2O_2, oxidation number of **H** has to be **+1** (an H atom can only lose one electron), so oxidation number of **O = -1**

8) Combined **hydrogen** is +1, except in metal hydrides where it is -1 (and H_2 where it's 0).

In **HF**, oxidation number of **H = +1**, but in **NaH**, oxidation number of **H = -1**

Roman Numerals Give Oxidation Numbers

Sometimes, oxidation numbers aren't clear from the formula of a compound.
If you see **Roman numerals** in a chemical name, it's an **oxidation number**.
E.g. copper has oxidation number **+2** in **copper(II) sulfate** and manganese has oxidation number **+7** in a **manganate(VII) ion** (MnO_4^-)

Hands up if you like Roman numerals...

Oxidation and Reduction

Oxidation Numbers go *Up* or *Down* as Electrons are *Lost* or *Gained*

1) The oxidation number for an atom will **increase by 1** for each **electron lost**.
2) The oxidation number will **decrease by 1** for each **electron gained**.
3) Elements can also be **oxidised and reduced** at the same time — this is called **disproportionation**.

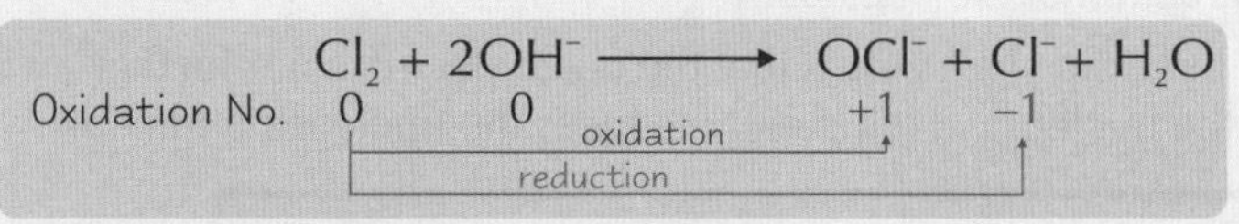

Example:
Chlorine and its ions undergo disproportionation reactions:

$$Cl_2 + 2OH^- \longrightarrow OCl^- + Cl^- + H_2O$$

Oxidation No. 0 0 → +1 (oxidation), −1 (reduction)

You can Separate Redox Reactions into *Half-Equations*

1) **Ionic half-equations** show oxidation or reduction.
2) You can **combine** half-equations for different oxidising or reducing agents together to make **full equations** for reactions.

Magnesium burns in oxygen to form magnesium oxide.
Magnesium is oxidised: $Mg \rightarrow Mg^{2+} + 2e^-$
Oxygen is reduced: $\frac{1}{2}O_2 + 2e^- \rightarrow O^{2-}$
Combining the half-equations gives: $Mg + \frac{1}{2}O_2 \rightarrow MgO$
(The electrons balance on each side so they aren't included in the full equation.)

Practice Questions

Q1 What is a reducing agent?

Q2 What is the usual oxidation number for oxygen combined with another element?

Q3 What is disproportionation?

Exam Questions

Q1 When hydrogen iodide gas is bubbled through warm concentrated sulfuric acid, hydrogen sulfide and iodine are produced.

a) Balance the equation below for the reaction.

$$H_2SO_{4\,(aq)} + HI_{(g)} \rightarrow H_2S_{(g)} + I_{2\,(s)} + H_2O_{(l)}$$ [1 mark]

b) State the oxidation number of sulfur in H_2SO_4 and in H_2S. [2 marks]

c) Write a half-equation to show the conversion of iodide, I^-, into iodine, I_2. [1 mark]

d) Write a half-equation to show the conversion of sulfuric acid into hydrogen sulfide. [2 marks]

e) In this reaction, which is the reducing agent? Give a reason. [2 marks]

Q2 If sodium hydroxide is added to aqueous solutions of iron(II) sulfate and iron(III) sulfate, precipitates of the corresponding hydroxides are formed.

a) Write the formulae for:

(i) iron(II) sulfate, (ii) iron(II) hydroxide, (iii) iron(III) sulfate, (iv) iron(III) hydroxide. [4 marks]

b) Most iron(II) compounds are green in colour but most iron(III) compounds are orange/brown. Suggest why iron-containing rocks that form in contact with air are more likely to be orange/brown. [1 mark]

Redox — relax in a lovely warm bubble bath...

Ionic equations are so evil even Satan wouldn't mess with them. But they're on the syllabus, so you can't ignore them. Have a flick back to p10 if they're freaking you out.

And while we're on the oxidation page, I suppose you ought to learn the most famous memory aid thingy in the world...

OIL RIG
- Oxidation Is Loss
- Reduction Is Gain
(of electrons)

Group 2

Group 2 is also known as the alkaline earth metals. They're in the "s block" of the periodic table and they're lovely.

Ionisation Energy ***Decreases*** *Down the Group*

1) Each element down Group 2 has an **extra electron shell** compared to the one above.
2) The extra inner shells **shield** the outer electrons from the attraction of the nucleus.
3) Also, the extra shell means that the outer electrons are **further away** from the nucleus, which greatly reduces the nucleus's attraction.

Both of these factors make it **easier** to remove outer electrons, resulting in a **lower ionisation energy**.

Mr Kelly has one final attempt at explaining electron shielding to his students...

The positive charge of the nucleus does increase as you go down a group (due to the extra protons), but this effect is overridden by the effect of the extra shells.

Group 2 Elements React with ***Water, Oxygen*** *and* ***Chlorine***

When Group 2 elements react, they're **oxidised** from a state of **0** to **+2**, forming M^{2+} ions. This is because Group 2 atoms contain 2 electrons in their outer shell.

M represents any Group 2 element.

	$M \rightarrow M^{2+} + 2e^-$	E.g.	$Ca \rightarrow Ca^{2+} + 2e^-$
Oxidation state:	**0** **+2**		**0** **+2**

1) **REACT WITH WATER TO PRODUCE HYDROXIDES**

The Group 2 metals react with water to give a **metal hydroxide and hydrogen**.
They get **increasingly** reactive down the group because the **ionisation energies** decrease.

$$M_{(s)} + 2H_2O_{(l)} \rightarrow M(OH)_{2\,(aq)} + H_{2\,(g)}$$

Oxidation state: M **0** → M(OH)₂ **+2**

e.g. $$Ca_{(s)} + 2H_2O_{(l)} \rightarrow Ca(OH)_{2\,(aq)} + H_{2\,(g)}$$

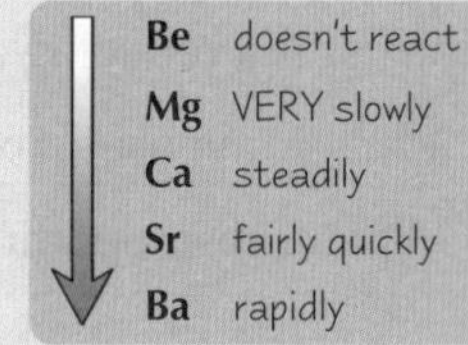

2) **BURN IN OXYGEN WITH CHARACTERISTIC FLAME COLOURS**

...to form solid white oxides.
Magnesium burns with a brilliant white flame, and the others burn with their characteristic **flame colours** (see p79).

$$2M_{(s)} + O_{2\,(g)} \rightarrow 2MO_{(s)}$$

Oxidation state of metal: 0 → +2
Oxidation state of oxygen: 0 → −2

e.g.

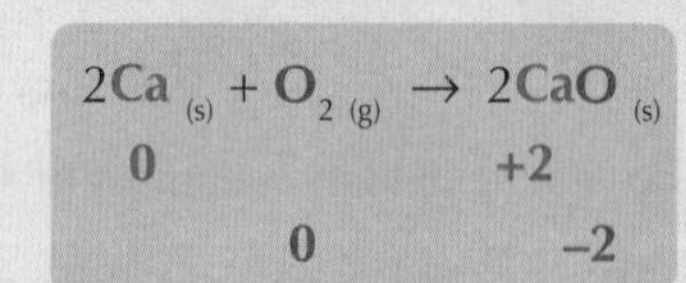

3) **REACTION WITH CHLORINE**

...forming white solid chlorides.

$$M_{(s)} + Cl_{2\,(g)} \rightarrow MCl_{2\,(s)}$$

Ox. state of metal: 0 → +2

e.g. $$Ca_{(s)} + Cl_{2\,(g)} \rightarrow CaCl_{2\,(s)}$$

0 → +2

The Oxides and Hydroxides are ***Bases***

1) The **oxides** of the Group 2 metals react readily with **water** to form **metal hydroxides**, which dissolve. The **hydroxide ions, OH^-**, make these solutions **strongly alkaline**.
2) Magnesium oxide is an exception — it only reacts slowly and the hydroxide isn't very soluble.
3) The oxides form **more strongly alkaline** solutions as you go down the group, because the hydroxides get more soluble.
4) Because they're **bases**, both the oxides and hydroxides will **neutralise** dilute acids, forming solutions of the corresponding salts. See the next page for examples of these reactions.

Group 2

...So They Form Alkaline Solutions and Neutralise Acids

	Reaction with Water	Reaction with Acid
Oxides	$MO_{(s)} + H_2O_{(l)} \rightarrow M(OH)_{2(aq)}$	$MO_{(s)} + 2HCl_{(aq)} \rightarrow MCl_{2(aq)} + H_2O_{(l)}$
Hydroxides	$M(OH)_{2(s)} \xrightarrow{+H_2O_{(l)}} M^{2+}_{(aq)} + 2OH^-_{(aq)}$	$M(OH)_{2(aq)} + 2HCl_{(aq)} \rightarrow MCl_{2(aq)} + 2H_2O_{(l)}$

Solubility Trends Depend on the Compound Anion

1) Generally, compounds of Group 2 elements that contain **singly charged** negative ions (e.g. OH^-) **increase** in solubility down the group...
2) ... whereas compounds that contain **doubly charged** negative ions (e.g. SO_4^{2-} and CO_3^{2-}) **decrease** in solubility down the group.
3) You need to know the solubility trends for the Group 2 **hydroxides** and the **sulfates**.

Group 2 element	hydroxide (OH^-)	sulfate (SO_4^{2-})
magnesium	least soluble ↓	most soluble ↑
calcium		
strontium		
barium	most soluble	least soluble

4) Most sulfates are soluble in water, but **barium sulfate** is **insoluble**. Compounds like magnesium hydroxide that have **very low** solubilities are said to be **sparingly soluble**.

Practice Questions

Q1 What happens to the ionisation energy as you move down Group 2?

Q2 Why does reactivity with water increase down Group 2?

Q3 Give one property of a Group 2 oxide that shows it to be a base.

Q4 Which is less soluble, barium sulfate or magnesium sulfate?

Exam Questions

Q1 Hydrochloric acid can be produced in excess quantities in the stomach, causing indigestion. Antacid tablets often contain sodium hydrogencarbonate which reacts with the acid to form a salt, carbon dioxide and water.

a) What discomfort could be caused by the carbon dioxide produced? [1 mark]

b) Name two magnesium compounds that could also be used as antacids and not cause discomfort. [1 mark]

c) Write an equation for its reaction with hydrochloric acid. [2 marks]

Q2 Calcium can be burned in chlorine gas.

a) Write an equation for the reaction. [1 mark]

b) Show the change in oxidation state of calcium. [1 mark]

c) Predict the appearance of the product. [2 marks]

d) What type of bonding would the product have? [1 mark]

Bored of Group 2 trends? Me too. Let's play noughts and crosses...

X	O	
X	O	X
O	O	

Noughts and crosses is pretty rubbish really, isn't it? It's always a draw. Ho hum. Back to Chemistry then, I guess...

Group 1 and 2 Compounds

These pages are about Groups 1 and 2, starting with the thermal stability of their carbonates and nitrates. So — quick, get your vest and long johns on before you topple over — we haven't even started yet.

Thermal Stability of Carbonates and Nitrates Changes Down the Group

Thermal decomposition is when a substance **breaks down** (decomposes) when **heated**.
The more thermally stable a substance is, the more heat it will take to break it down.

1) **Thermal stability increases down a group**

   The carbonate and nitrate ions are **large** and can be made **unstable** by the presence of a **positively charged ion** (a cation). The cation **polarises** the anion, distorting it. The greater the distortion, the less stable the anion.

   Large cations cause **less distortion** than small cations. So the further down the group, the larger the cations, the less distortion caused and the **more stable** the carbonate/nitrate anion. Phew... that was hard.

2) **Group 2 compounds are less thermally stable than Group 1 compounds**

   The greater the **charge** on the cation, the greater the **distortion** and the **less stable** the carbonate/nitrate ion becomes. Group 2 cations have a **2+** charge, compared to a **1+** charge for Group 1 cations.
   So Group 2 carbonates and nitrates are less stable than those of Group 1.

Group 1	Group 2
Group 1 carbonates* are **thermally stable** — you can't heat them enough with a Bunsen to make them decompose (though they do decompose at higher temperatures). *except Li_2CO_3 which decomposes to Li_2O and CO_2 (there's always one...)	Group 2 carbonates decompose to form the **oxide** and **carbon dioxide**. $MCO_{3\,(s)} \rightarrow MO_{(s)} + CO_{2\,(g)}$ e.g. $CaCO_{3\,(s)} \rightarrow CaO_{(s)} + CO_{2\,(g)}$ calcium carbonate → calcium oxide
Group 1 nitrates** decompose to form the **nitrite** and **oxygen**. $2MNO_{3\,(s)} \rightarrow 2MNO_{2\,(s)} + O_{2\,(g)}$ e.g. $2KNO_{3\,(s)} \rightarrow 2KNO_{2\,(s)} + O_{2\,(g)}$ potassium nitrate → potassium nitrite **except $LiNO_3$ which decomposes to form Li_2O, NO_2 and O_2.	Group 2 nitrates decompose to form the **oxide**, **nitrogen dioxide** and **oxygen**. $2M(NO_3)_{2\,(s)} \rightarrow 2MO_{(s)} + 4NO_{2\,(g)} + O_{2\,(g)}$ e.g. $2Ca(NO_3)_{2\,(s)} \rightarrow 2CaO_{(s)} + 4NO_{2\,(g)} + O_{2\,(g)}$ calcium nitrate → calcium oxide, nitrogen dioxide

Here's How to Test the Thermal Stability of Nitrates and Carbonates

How easily nitrates decompose can be tested by measuring...
- how long it takes until **oxygen** is produced (i.e. to relight a glowing splint)

OR
- how long it takes until a **brown gas (NO_2)** is produced.
  This needs to be done in a fume cupboard because NO_2 is **toxic**.

Daisy the cow *

How easily carbonates decompose can be tested by measuring...
- how long it takes for **carbon dioxide** to be produced.
  You test for carbon dioxide using lime water — which is a saturated solution of calcium hydroxide. This turns cloudy with carbon dioxide.

* She wanted to be in the book. I said OK.

Group 1 and 2 Compounds

Group 1 and 2 Compounds Burn with Distinctive **Flame Colours**

...not all of them, but quite a few. For compounds containing the ions below, flame tests can help **identify them**.

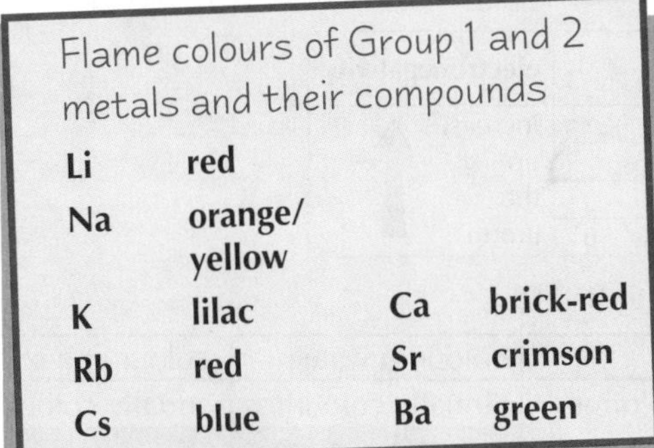

Flame colours of Group 1 and 2 metals and their compounds

Li	**red**		
Na	**orange/ yellow**		
K	**lilac**	**Ca**	**brick-red**
Rb	**red**	**Sr**	**crimson**
Cs	**blue**	**Ba**	**green**

Here's how to do a flame test:

1) Mix a small amount of the compound you're testing with a few drops of **hydrochloric acid**.
2) Heat a piece of **platinum** or **nichrome wire** in a hot Bunsen flame to clean it.
3) Dip the wire into the compound/acid mixture. Hold it in a very hot flame and note the colour produced.

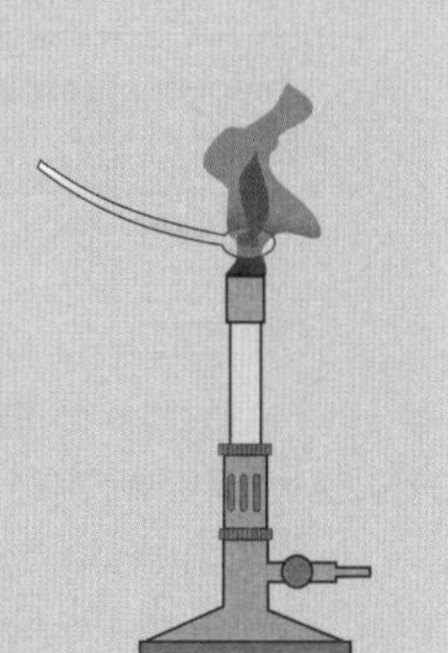

The explanation

The **energy** absorbed from the flame causes electrons to move to **higher energy levels**. The colours are seen as the electrons fall back down to lower energy levels, releasing energy in the form of **light**. The difference in energy between the higher and lower levels determines the **wavelength** of the light released — which determines the **colour** of the light.

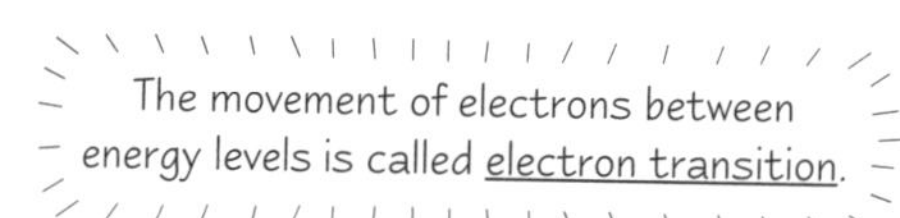

Practice Questions

Q1 Write an equation for the thermal decomposition of calcium carbonate.

Q2 What is the trend in the thermal stability of the nitrates of Group 1 elements?

Q3 Describe two ways that you could test how easily the nitrates of Group 2 decompose.

Q4 Which Group 1 or 2 metal ions are indicated by the following flame colours?
a) lilac b) brick-red c) orange/yellow

Exam Questions

Q1 When heated, a compound of calcium produces a gas **A** and a solid **B**.
The gas **A** is bubbled through a solution of limewater to give a cloudy precipitate.
Give the formulae of the substances **A** and **B**, and the formula of the original compound. [3 marks]

Q2 a) Write a balanced equation for the thermal decomposition of sodium nitrate. [1 mark]

b) How could you test for the gas produced in the thermal decomposition? [1 mark]

c) Place the following in order of ease of thermal decomposition (easiest first).

magnesium nitrate **potassium nitrate** **sodium nitrate**

Explain your answer. [3 marks]

Q3 a) When a substance is heated, what changes occur within the atom that give rise to a coloured flame? [2 marks]

b) A compound gives a blue colour in a flame test.
What s-block metal ions might this compound contain? [1 mark]

"So that was lithium, now let's try..." "SCHTOP! — this flame is not reddy yet..."

Here at CGP, we like to test our flames slowly. So they burn real smooth with their characteristic colours...
[OK — Group 1 and 2 compounds. Lots of trends here. Just learn them. And learn the explanation for each one (not like a parrot – so you understand it). Then learn the flame colours and that explainy bit about energy levels, electron transitions and wavelengths.] CGP — we only test flames when they're ready...

The Halogens

Here comes a page jam-packed with golden nuggets of halogen fun. Oh yes, I kid you not. This page is the Alton Towers of AS Chemistry... white-knuckle excitement all the way...

The word halogen should be used when describing the atom (X) or molecule (X_2), but the word halide is used to describe the negative ion (X^-).

Halogens are the **Highly Reactive Non-Metals** of Group 7

1) The table below gives some of the main properties of the first 4 halogens.

halogen	formula	colour	physical state	electronic structure	electronegativity
fluorine	F_2	pale yellow	gas	$1s^2\ 2s^2\ 2p^5$	increases
chlorine	Cl_2	green	gas	$1s^2\ 2s^2\ 2p^6\ 3s^2\ 3p^5$	up
bromine	Br_2	red-brown	liquid	$1s^2\ 2s^2\ 2p^6\ 3s^2\ 3p^6\ 3d^{10}\ 4s^2\ 4p^5$	the
iodine	I_2	grey	solid	$1s^2\ 2s^2\ 2p^6\ 3s^2\ 3p^6\ 3d^{10}\ 4s^2\ 4p^6\ 4d^{10}\ 5s^2\ 5p^5$	group

2) Halogens in their natural state exist as covalent diatomic molecules (e.g. Br_2, Cl_2). Because they're covalent, they have **low solubility in water**.
But they do dissolve easily in **organic compounds** like hexane. Some of these resulting solutions have distinctive colours that can be used to identify them.

	colour in water	colour in hexane
chlorine	virtually colourless	virtually colourless
bromine	yellow/orange	orange/red
iodine	brown	pink/violet

Halogens get **Less Reactive** Down the Group

1) Halogen atoms react by **gaining an electron** in their outer p sub-shell. This means they're **reduced**. As they're reduced, they **oxidise** another substance (it's a redox reaction) — so they're **oxidising agents**.

$$X + e^- \rightarrow X^-$$
ox. state: 0 → -1

2) As you go down the group, the atoms become **larger** so the outer electrons are **further** from the nucleus. The outer electrons are also **shielded** more from the attraction of the positive nucleus, because there are more inner electrons. This makes it harder for larger atoms to attract the electron needed to form an ion, so larger atoms are less reactive.
3) Another way of saying that the halogens get **less reactive** down the group is to say that they become **less oxidising**. (See the next page for more on this.)

You Can Use **Patterns** to **Predict Properties**

1) The smallest halogen, **fluorine**, is the **most reactive** non-metal element. Fluorine isn't used in schools and colleges because it's so dangerous, but you can **predict** its properties by looking at those of the other halogens.
2) The **melting** and **boiling points** increase down the group, so you can predict that fluorine would be a gas at room temperature, like chlorine below it. Similarly, fluorine should be **coloured** as all the other halogens are. In fact, fluorine is a very pale yellow gas at room temperature.
3) Astatine (below iodine in the periodic table) is a solid. You'd expect it to be the **least reactive** halogen, but its properties haven't been studied because it's highly radioactive and decays quickly.

Halogen	Melting Point / °C	Boiling Point / °C
F		
Cl	–101	–34
Br	–7	58
I	114	183
At		

Increasing Reactivity (arrow pointing up from At to F)

Halogens undergo **Disproportionation** with Alkalis

The halogens react with hot and cold alkali solutions. In these reactions, the halogen is simultaneously oxidised and reduced (called **disproportionation**)...

You can use other alkalis too, e.g. KOH.

COLD

$$X_2 + 2NaOH \rightarrow NaXO + NaX + H_2O$$

Ionic equation: $X_2 + 2OH^- \rightarrow XO^- + X^- + H_2O$

Ox. state of X: 0 → +1, –1

Example: $I_2 + 2NaOH \rightarrow NaIO + NaI + H_2O$ (NaIO = sodium iodate(I))

HOT

$$3X_2 + 6NaOH \rightarrow NaXO_3 + 5NaX + 3H_2O$$

Ionic equation: $3X_2 + 6OH^- \rightarrow XO_3^- + 5X^- + 3H_2O$

Ox. state of X: 0 → +5, –1

Example: $3Br_2 + 6NaOH \rightarrow NaBrO_3 + 5NaBr + 3H_2O$ ($NaBrO_3$ = sodium bromate(III))

The halogens (except fluorine) can exist in a wide range of oxidation states e.g.

-1	0	+1	+1	+3	+5	+7
Cl^-	Cl_2	ClO^-	BrO^-	BrO_2^-	IO_3^-	IO_4^-
chloride	chlorine	chlorate(I)	bromate(I)	bromate(III)	iodate(V)	iodate(VII)

All the **halogen -ate ions** have a **single halogen atom** and a charge of **–1** — so if you forget the formula you should be able to work it out from the **oxidation number**.

The Halogens

Here's Some More Reactions of the Halogens to **Learn**...

Remember, when halogens react they're reduced — and they oxidise other substances.

They Oxidise Metals...

For example, **fluorine** and **chlorine** react with hot **iron** to form iron(III) halides. (Iron is taken to its highest oxidation state, +3, because these halogens are very **strong oxidising agents**.)

$$2Fe_{(s)} + 3Cl_{2\,(g)} \rightarrow 2FeCl_{3\,(s)}$$

The iron(III) chloride is produced as a vapour, which condenses to a solid.

Iron is oxidised: $2Fe \rightarrow 2Fe^{3+} + 6e^-$

Chlorine is reduced: $3Cl_2 + 6e^- \rightarrow 6Cl^-$

Bromine's a **weaker** oxidising agent so you get a mixture of iron(**II**) and iron(**III**) bromide. With **iodine**, you only get iron(**II**) iodide — no Fe^{3+} ions form.

...and Non-Metals

For example, chlorine reacts with **sulfur** to form sulfur(I) chloride. (Sulfur is oxidised to +1 and chlorine is reduced to –1.)

$$S_{8\,(s)} + 4Cl_{2\,(g)} \rightarrow 4S_2Cl_{2\,(l)}$$

...and Some Ions

For example, all the halogens except iodine (which is less strongly oxidising than the others) will oxidise iron(II) ions to iron(III) ions in solution. The solution will change colour from **green** to **orange**.

For example: $Cl_2 + 2e^- \rightarrow 2Cl^-_{(aq)}$

$$2Fe^{2+}_{(aq)} \rightarrow 2Fe^{3+}_{(aq)} + 2e^-$$

green (Fe^{2+}) **orange** (Fe^{3+})

Practice Questions

Q1 What colour is a solution of bromine in water? And in hexane?

Q2 What does disproportionation mean?

Q3 Write the ionic equation for the reaction of hot potassium hydroxide with iodine.

Exam Questions

Q1 If chlorine gas and sodium hydroxide are allowed to mix at room temperature, sodium chlorate(I) is formed.

a) Give the ionic equation for the reaction. [1 mark]

b) This is a disproportionation reaction.
Use the ionic equation for this reaction to explain what is meant by disproportionation. [3 marks]

Q2 Write formulae for the following compounds (assume all the anions have a charge of -1).

a) magnesium fluoride b) potassium bromate(I) c) sodium chlorate(V) [3 marks]

Q3 When copper foil is heated in a stream of dry chlorine, the copper is oxidised to its highest oxidation state and a white powder is formed as the only product.

a) Name the powder produced. [1 mark]

b) Write half-equations for the oxidation and reduction steps in this reaction.
Indicate which equation shows oxidation and which shows reduction. [3 marks]

Remain seated until the page comes to a halt. Please exit to the right...

Oooh, what a lovely page, if I do say so myself. I bet the question of how halogens react with hot and cold alkali has plagued your mind since childhood. Well now you know. There's nowt too taxing here — you just need to learn the colours of the solutions, all the equations, and make sure you can tell what's oxidised and what's reduced... it never ends.

Reactions of the Halides

Ah, halides. Personally, I can never get enough of them.

The **Reducing Power** of Halides **Increases** Down the Group...

A halide ion can act as a **reducing agent** by losing an electron from its outer shell. (See the reactions with halogens on the next page.) How easy this is depends on the **attraction** between the halide's **nucleus** and the outer **electrons**. As you go down the group, the attraction gets **weaker** because:

1) the ions get bigger, so the electrons are **further** away from the positive nucleus
2) there are extra inner electron shells, so there's a greater **shielding** effect.

...which Explains their Reactions with **Sulfuric Acid**

All the halides react with concentrated sulfuric acid to give a **hydrogen halide** as a product to start with. But what happens next depends on which halide you've got...

Reaction of KF or KCl with H_2SO_4

$$KF_{(s)} + H_2SO_{4(l)} \rightarrow KHSO_{4(s)} + HF_{(g)}$$

$$KCl_{(s)} + H_2SO_{4(l)} \rightarrow KHSO_{4(s)} + HCl_{(g)}$$

1) Hydrogen fluoride (HF) or hydrogen chloride gas (HCl) is formed. You'll see misty fumes as the gas comes into contact with moisture in the air.
2) But HF and HCl aren't strong enough reducing agents to reduce the sulfuric acid, so the reaction stops there.
3) It's not a redox reaction — the oxidation states of the halide and sulfur stay the same (–1 and +6).

Reaction of KBr with H_2SO_4

$$KBr_{(s)} + H_2SO_{4(l)} \rightarrow KHSO_{4(s)} + HBr_{(g)}$$

$$2HBr_{(aq)} + H_2SO_{4(l)} \rightarrow Br_{2(g)} + SO_{2(g)} + 2H_2O_{(l)}$$

ox. state of S: +6 → +4 reduction

ox. state of Br: -1 → 0 oxidation

1) The first reaction gives misty fumes of hydrogen bromide gas (HBr).
2) But the HBr is a stronger reducing agent than HCl and reacts with the H_2SO_4 in a redox reaction.
3) The reaction produces choking fumes of SO_2 and orange fumes of Br_2.

Reaction of KI with H_2SO_4

$$KI_{(s)} + H_2SO_{4(l)} \rightarrow KHSO_{4(s)} + HI_{(g)}$$

$$2HI_{(g)} + H_2SO_{4(l)} \rightarrow I_{2(s)} + SO_{2(g)} + 2H_2O_{(l)}$$

ox. state of S: +6 → +4 reduction

ox. state of I: -1 → 0 oxidation

$$6HI_{(g)} + SO_{2(g)} \rightarrow H_2S_{(g)} + 3I_{2(s)} + 2H_2O_{(l)}$$

ox. state of S: +4 → –2 reduction

ox. state of I: -1 → 0 oxidation

1) Same initial reaction giving HI gas.
2) The HI then reduces H_2SO_4 as above.
3) But HI (being well 'ard as far as reducing agents go) keeps going and reduces the SO_2 to H_2S.

H_2S gas is toxic and smells of bad eggs. A bit like my mate Andy at times...

Hydrogen Halides are **Acidic Gases**

The **hydrogen halides** are **colourless gases**, but you can't forget about them just cos you can't see 'em.

1) They're **very soluble**, dissolving in water to make **strong acids**. (They'll happily turn blue litmus paper red.) → $HCl_{(g)} \rightarrow H^+_{(aq)} + Cl^-_{(aq)}$
2) Hydrogen chloride forms **hydrochloric** acid, hydrogen bromide forms **hydrobromic** acid and hydrogen iodide gives **hydroiodic** acid. (You don't hear of this last one much — that's because its name is too silly.)
3) They react with **ammonia gas** to give **white fumes**. E.g. hydrogen chloride gives ammonium chloride. → $NH_{3(g)} + HCl_{(g)} \rightarrow NH_4Cl_{(s)}$ (It's an acid-base reaction.)

Reactions of the Halides

Halide Ions Are **Displaced** from Solution by **More Reactive** Halogens

1) The halogens' **relative oxidising strengths** can be seen in their **displacement reactions** with halide ions. For example, if you mix bromine water, $Br_{2\,(aq)}$, with potassium iodide solution, the bromine displaces the iodide ions (it oxidises them), giving iodine, $I_{2\,(aq)}$ and potassium bromide, $KBr_{(aq)}$. You can see what happens by following the **colour changes**.

	Potassium chloride solution **$KCl_{(aq)}$** - colourless	Potassium bromide solution **$KBr_{(aq)}$** - colourless	Potassium iodide solution **$KI_{(aq)}$** - colourless
Chlorine water **$Cl_{2(aq)}$** - colourless	no reaction	orange solution (Br_2) formed	brown solution (I_2) formed
Bromine water **$Br_{2(aq)}$** - orange	no reaction	no reaction	brown solution (I_2) formed
Iodine solution **$I_{2(aq)}$** - brown	no reaction	no reaction	no reaction

2) You can make the changes easier to see by shaking the reaction mixture with an **organic solvent** like hexane. The halogen that's present will dissolve readily in the organic solvent, which settles out as a distinct layer above the aqueous solution. This example shows the presence of **iodine**.
(The colours of the other solutions are on p80.)

hexane layer
aqueous layer

3) These displacement reactions can be used to help **identify** which halogen (or halide) is present in a solution.

A **halogen** will **displace a halide** from solution if the halide is **below it** in the periodic table, e.g.

Halogen	Displacement reaction	Ionic equation
Cl	chlorine (Cl_2) will displace bromide (Br^-) and iodide (I^-)	$Cl_{2(aq)} + 2Br^-_{(aq)} \rightarrow 2Cl^-_{(aq)} + Br_{2(aq)}$ $Cl_{2(aq)} + 2I^-_{(aq)} \rightarrow 2Cl^-_{(aq)} + I_{2(aq)}$
Br	bromine (Br_2) will displace iodide (I^-)	$Br_{2(aq)} + 2I^-_{(aq)} \rightarrow 2Br^-_{(aq)} + I_{2(aq)}$
I	no reaction with F^-, Cl^-, Br^-	

You can also say a halogen will **oxidise** a halide if the halide is below it in the periodic table.

$$Cl_{2(aq)} + 2Br^-_{(aq)} \rightarrow 2Cl^-_{(aq)} + Br_{2(aq)}$$

ox. state of Cl	**0**		→	**-1**	reduction
ox. state of Br		**-1**	→	**0**	oxidation

Halides Give **Coloured Precipitates** with **Silver Nitrate Solution**

Here's a nice **test** to help you find out which **halide ions** you're dealing with:

1) First you add **dilute nitric acid** to remove ions that might interfere with the test.
2) Then you just add **silver nitrate solution** ($AgNO_{3(aq)}$).
   A **precipitate** is formed (of the silver halide).

$$Ag^+_{(aq)} + X^-_{(aq)} \rightarrow AgX_{(s)} \quad \text{...where X is F, Cl, Br or I}$$

The **colour** of the precipitate identifies the halide.
Then to be extra sure, you can test your results by adding **ammonia solution**.
(Each silver halide has a different solubility in ammonia.)

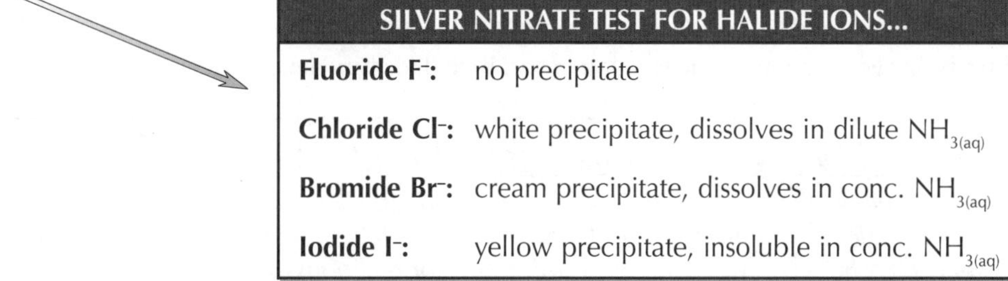

SILVER NITRATE TEST FOR HALIDE IONS...	
Fluoride F^-:	no precipitate
Chloride Cl^-:	white precipitate, dissolves in dilute $NH_{3(aq)}$
Bromide Br^-:	cream precipitate, dissolves in conc. $NH_{3(aq)}$
Iodide I^-:	yellow precipitate, insoluble in conc. $NH_{3(aq)}$

Reactions of the Halides

Silver Halides React with Sunlight

Silver halides **decompose** when light shines on them, producing **silver** and the **halogen**. For example:

$$2AgBr \rightarrow 2Ag + Br_2$$

This reaction is used in film photography — the film contains silver bromide particles that turn to opaque silver when they're exposed to light.

You Might Have to Make Predictions

1) As I'm sure you know by now, halide ions are **reducing agents**, and they get more **strongly** reducing as you go down the periodic table (see p82).

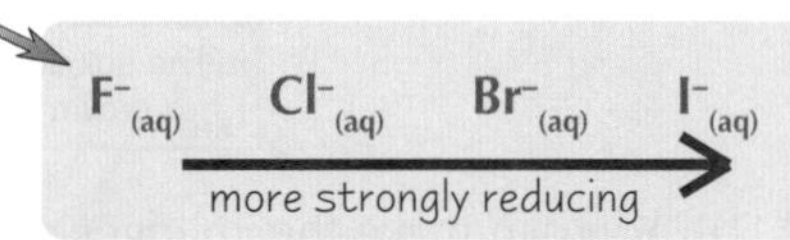

2) For example, iron(III) iodide solution is unstable, because **iodide** ions are strong enough reducing agents to reduce the Fe^{3+} ions to Fe^{2+} ions:

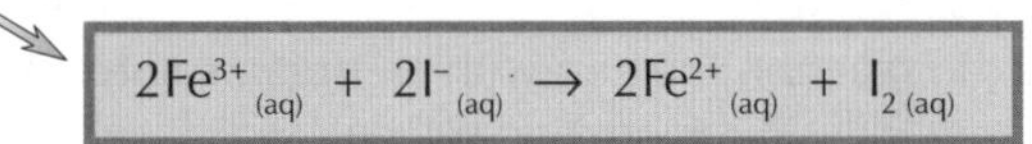

3) Iron(III) **chloride** however, is stable because Cl^- ions aren't powerful enough reducing agents to reduce Fe^{3+} ions.
4) You can use patterns like this to make predictions. For example, now you know that iodide ions will reduce Fe^{3+}, you could predict that **astatide** ions, At^- would do the same (At is below I in the periodic table, so At– would be even more strongly reducing)... and so iron(III) astatide solution will also be unstable.

Practice Questions

Q1 Give two reasons why a bromide ion is a more powerful reducing agent than a chloride ion.

Q2 Name the gaseous products formed when potassium bromide reacts with concentrated sulfuric acid.

Q3 What do you see when potassium iodide reacts with concentrated sulfuric acid?

Q4 What type of substance is formed when a hydrogen halide is passed through water?

Q5 What would you see if you mixed hydrogen iodide with ammonia?

Q6 What would you observe when bromine water is mixed with potassium chloride solution?

Exam Questions

Q1 Describe the tests you would carry out in order to distinguish between solid samples of potassium chloride and potassium bromide using: a) silver nitrate solution and aqueous ammonia,
b) concentrated sulfuric acid.
For each test, state your observations and write an equation for the reaction which occurs. [11 marks]

Q2 The halogen below iodine in Group 7 is astatine (At). Predict, giving an explanation, whether or not:
a) hydrogen sulfide gas would be evolved when concentrated sulfuric acid is added to a solid sample of potassium astatide [4 marks]
b) silver astatide will dissolve in concentrated ammonia solution [3 marks]

Q3 Look at the following equation:

$$2Cu^{2+}_{(aq)} + 4I^-_{(aq)} \rightarrow 2CuI_{(s)} + I_{2\,(aq)}$$

a) Explain why this reaction can be referred to as a redox reaction, and identify the reducing agent. [2 marks]
b) Describe what you would observe when chlorine gas is bubbled through a solution of potassium iodide. [2 mark]
c) Write a balanced equation for the reaction in part b) and identify the oxidising agent. [3 marks]

[Sing along with me] "Why won't this section end... Why won't this section end..."

AS Chemistry. What a bummer, eh... No one ever said it was going to be easy. Not even your teacher would be that cruel. There are plenty more equations on these pages to learn. As well as that, make sure you really understand everything... what exactly reducing agents do... how you work out oxidation states for reactions... And no, you can't swap to English. Sorry.

Acid-Base Titrations and Uncertainty

*Titrations are used to find out the **concentration** of acid or alkali solutions. They're also handy when you're making soluble salts of soluble bases.*

Titrations need to be done Accurately

1) **Titrations** allow you to find out **exactly** how much acid is needed to **neutralise** a quantity of alkali.
2) You measure out some **alkali** using a pipette and put it in a flask, along with some **indicator**, e.g. **phenolphthalein**.
3) First of all, do a rough titration to get an idea where the **end point** is (the point where the alkali is **exactly neutralised** and the indicator changes colour). Add the **acid** to the alkali using a **burette** — giving the flask a regular **swirl**.
4) Now do an **accurate** titration. Run the acid in to within 2 cm^3 of the end point, then add the acid **dropwise**. If you don't notice exactly when the solution changed colour you've **overshot** and your result won't be accurate.
5) **Record** the amount of acid used to **neutralise** the alkali. It's best to **repeat** this process a few times, making sure you get the same answer each time.

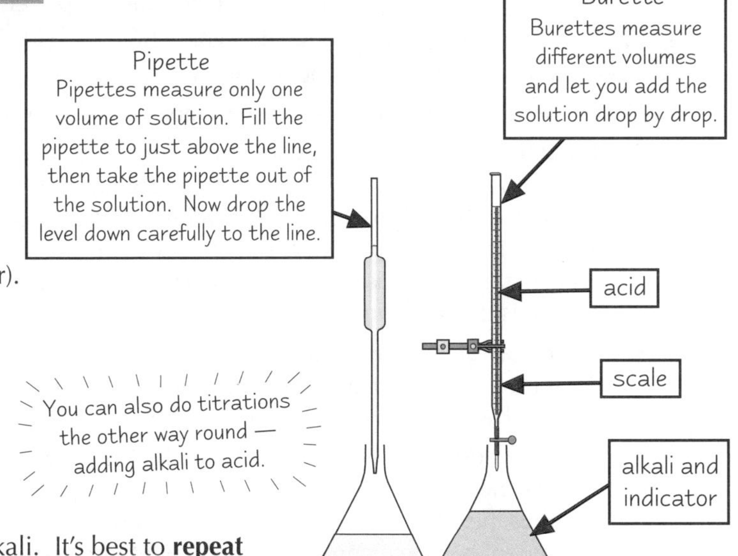

Indicators Show you when the Reaction's Just Finished

Indicators change **colour**, as if by magic. In titrations, indicators that change colour quickly over a **very small pH range** are used so you know **exactly** when the reaction has ended.

The main two indicators for **acid/alkali reactions** are —

> **methyl orange** — turns **yellow** to **red** when adding acid to alkali.
> **phenolphthalein** — turns **red** to **colourless** when adding acid to alkali.

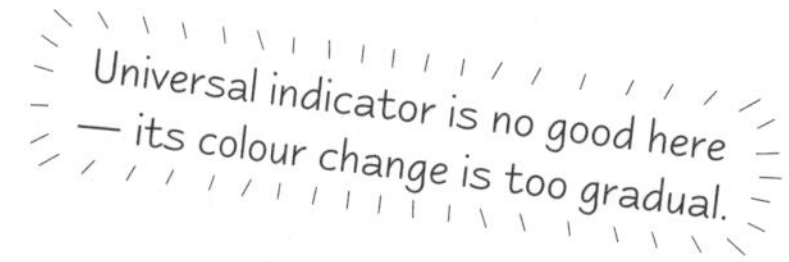

You can Calculate Concentrations from Titrations

Example: 25 cm^3 of 0.5 M HCl was used to neutralise 35 cm^3 of NaOH solution. Calculate the concentration of the sodium hydroxide solution.

First write a **balanced equation** and decide **what you know** and what you **need to know**:

$$HCl + NaOH \rightarrow NaCl + H_2O$$

HCl	NaOH
25 cm^3	**35 cm^3**
0.5 M	**?**

It's just the formula from page 7.

$$\text{Number of moles HCl} = \frac{\text{concentration} \times \text{volume (cm}^3)}{1000} = \frac{0.5 \times 25}{1000} = 0.0125 \text{ moles}$$

From the equation, you know 1 mole of HCl neutralises 1 mole of NaOH.
So 0.0125 moles of HCl must neutralise **0.0125** moles of NaOH.

Now it's a doddle to work out the **concentration of NaOH**.

$$\text{Concentration of NaOH}_{(aq)} = \frac{\text{moles of NaOH} \times 1000}{\text{volume (cm}^3)} = \frac{0.0125 \times 1000}{35} = \mathbf{0.36 \text{ mol dm}^{-3}}$$

If you're asked for the concentration in g dm^{-3}, you need to now use the formula from p6 — number of moles = mass ÷ M_r

Acid-Base Titrations and Uncertainty

Thought you were done with titrations? Well they're not finished with you yet...

Uncertainty is the Amount of Error Your Measurements Might Have

The results you get from a titration won't be completely perfect.

1) When you do a **titration**, you need to know how to work out how much **error** there could be in your **measurements**.
2) The **maximum possible error** is a useful measure of **uncertainty**.
   - The **uncertainty** in your measurements **varies** for different equipment. For example, the scale on a 50 cm^3 **burette** has marks every **0.1 cm^3**. You should be able to tell which mark the level's closest to, so any reading you take won't be more than **0.05 cm^3** out (as long as you don't make a daft mistake). The **uncertainty** of a reading from the burette is the **maximum error** you could have — so that's **0.05 cm^3**.
   - There's **uncertainty** when you weigh stuff, too. Even electronic scales don't give an **exact mass**. If the mass is measured to the **nearest 0.01 g**, the real mass could be up to **0.005 g smaller or larger**.
   - Pieces of equipment for measuring out **liquid** — things like fixed-volume pipettes and volumetric flasks — have uncertainties in the **volumes** they measure. These depend on how well made the equipment is. The manufacturers provide these **uncertainty values**.

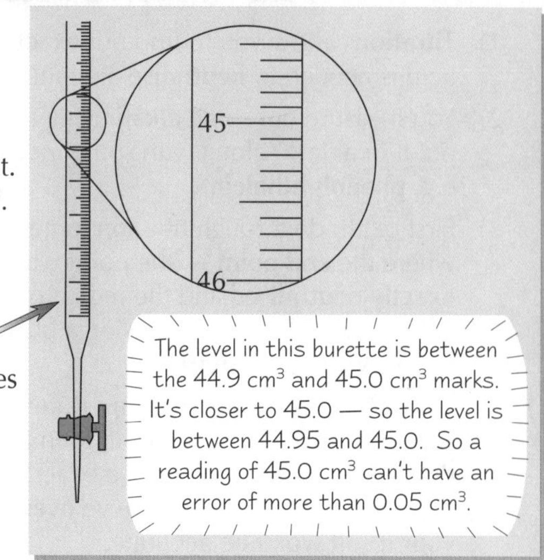

The level in this burette is between the 44.9 cm^3 and 45.0 cm^3 marks. It's closer to 45.0 — so the level is between 44.95 and 45.0. So a reading of 45.0 cm^3 can't have an error of more than 0.05 cm^3.

You Can Minimise Some Uncertainties

1) One obvious way to **reduce errors** in your measurements is to buy the most **precise equipment** available. In real life there's not much you can do about this one — you're stuck with whatever your school or college has got. But there are other ways to **lower the uncertainty** in your titrations.
2) A bit of clever **planning** can improve your results. Think about the readings from a **burette**. You take **two readings** to work out a titre (the volume of liquid delivered from the burette) — the **initial volume** and the **final volume**. Each reading has an uncertainty of **0.05 cm^3**. The titre is the second reading minus the first, so the titre will have a total uncertainty of **0.1 cm^3**. (The second could be up to 0.05 cm^3 too high, and the first up to 0.05 cm^3 too low.)
3) For any reading or measurement you can calculate the **percentage uncertainty** using the equation:

$$\text{percentage uncertainty} = \frac{\text{uncertainty}}{\text{reading}} \times 100$$

Percentage uncertainty is sometimes called percentage error.

4) If you use a burette to measure **10 cm^3** of liquid the percentage uncertainty is $(0.1/10) \times 100 =$ **1%**. But if you measure **20 cm^3** of liquid the uncertainty is $(0.1/20) \times 100 =$ **0.5%**. Hey presto — you've just halved the uncertainty. The percentage uncertainty can be reduced by planning a titration so that a **larger volume** will be measured by the burette.
5) The same principle can be applied to other measurements such as **weighing solids** — if you plan to weigh a small mass, the **percentage uncertainty** will be large.

Errors Can Be Systematic or Random

1) **Systematic errors** are the same every time you repeat the experiment. They may be caused by the **set-up** or **equipment** you're using. If the 10.00 cm^3 pipette you're using to measure out a sample for titration actually only measures 9.95 cm^3, your sample will be about 0.05 cm^3 too small **every time** you repeat the experiment.
2) **Random errors** vary — they're what make the results a bit **different** each time you repeat an experiment. The errors when you make a reading from a burette are random. You have to estimate or round the level when it's between two marks — so sometimes your figure will be **above** the real one, and sometimes it will be **below**.
3) **Repeating an experiment** and finding the mean of your results helps to deal with **random errors**. The results that are are bit high will be **cancelled out** by the ones that are a bit low. (Your results will be more **reliable**.) But repeating your results won't get rid of any **systematic errors**. (Your results won't get more **accurate**.)

This should be a photo of a scientist. I don't know what happened — it's a random error...

Acid-Base Titrations and Uncertainty

The *Total Uncertainty* in a Result Should be Calculated

1) In chemical analysis, knowing the **uncertainty** in the **final result** can be really important. For instance, if you were analysing the alcohol level of a driver's blood and found that it was just above the legal limit, this data would be no use to the police if the uncertainty was large enough that the driver could have been just under the limit.
2) In **titrations**, here's how you find the **total uncertainty in the final result**:
   - Find the **percentage uncertainty** for each bit of equipment.
   - Add the individual percentage uncertainties together. This gives the **percentage uncertainty in the final result**.
   - Use this to work out the **actual total uncertainty** in the final result.

Example: 10.00 cm^3 of KOH solution is neutralised by 27.3 cm^3 of HCl of known concentration.
The volume of KOH has an uncertainty of 0.06 cm^3. The volume of HCl has an uncertainty of 0.1 cm^3.
The concentration of the KOH is calculated to be 1.365 mol dm^{-3}.
What is the uncertainty in this concentration?

First work out the **percentage uncertainty** for each **volume measurement**:

The KOH volume of 10.00 cm^3 has an uncertainty of 0.06 cm^3:

$$\text{percentage uncertainty} = \frac{0.06}{10.00} \times 100 = \mathbf{0.60\%}$$

The HCl volume of 27.3 cm^3 has an uncertainty of 0.1 cm^3:

$$\text{percentage uncertainty} = \frac{0.1}{27.3} \times 100 = \mathbf{0.37\%}$$

Find the **percentage uncertainty in the final result:** Total percentage uncertainty = 0.60% + 0.37% = **0.97%**

You're not done yet — you still have to calculate the **uncertainty** in the final result.

Uncertainty in the final answer is 0.97% of 1.365 mol dm^{-3} = **0.013 mol dm^{-3}**

So the actual concentration may be 0.013 mol dm^{-3} bigger or smaller than 1.365 mol dm^{-3}.

Practice Questions

Q1 If the uncertainty of a reading from a burette is 0.05 cm^3, why is the uncertainty of a titre quoted as being 0.1 cm^3?

Q2 Write down the equation for the percentage uncertainty of a measurement.

Q3 Does repeating the same experiment several times improve the reliability or the accuracy of the results?

Exam Questions

Run	Initial volume (cm^3)	Final volume (cm^3)	Titre (cm^3)
Rough	1.1	5.2	4.1
1	1.2	4.3	3.1

Q1 The table shows the data recorded from a titration experiment.

a) Suggest a way to make the data more reliable. [1 mark]

b) Each reading recorded in the experiment has an uncertainty of 0.05 cm^3.
Calculate the percentage uncertainty in the **titre** in Run 1. [2 marks]

c) Explain how you could reduce the percentage error in these titre values by changing the concentration of the solution in the burette. [2 marks]

Q2 The concentration of a solution of sodium hydroxide is measured by titration against 0.100 M hydrochloric acid.
25.00 cm^3 of NaOH solution requires 19.25 cm^3 of HCl for neutralisation.

a) Calculate the concentration of the NaOH. [3 marks]

b) The volume of NaOH was measured using a pipette with an uncertainty of 0.06 cm^3.
The titre reading from the burette has an uncertainty of 0.1 cm^3.
By combining percentage uncertainties calculate the uncertainty in the concentration of the NaOH. [4 marks]

I used to be uncertain, but now I'm not sure...

Typical... you think you've done a nice, accurate experiment and all they care about is how wrong it is. If you get a question about uncertainty, make sure you read it carefully. If the question asks for the uncertainty in the final answer, they want the uncertainty in the same units as the result. If you only work out the total percentage uncertainty, you'll miss out.

Iodine-Sodium Thiosulfate Titration

Titration calculations get a wee bit more complicated than the one on page 85...

Iodine-Sodium Thiosulfate Titrations are Dead Handy

Iodine-sodium thiosulfate titrations are a way of finding the concentration of an **oxidising agent**.
The **more concentrated** an oxidising agent is, the **more ions will be oxidised** by a certain volume of it.
So here's how you can find out the concentration of a solution of the oxidising agent **potassium iodate(V)**:

STAGE 1: Use a sample of oxidising agent to oxidise as much iodide as possible.

1) Measure out a certain volume of potassium iodate(V) (the oxidising agent) — say **25 cm³**.
2) Add this to an excess of acidic **potassium iodide** solution. The iodate(V) ions in the potassium iodate(V) solution **oxidise** some of the **iodide ions** to **iodine**.

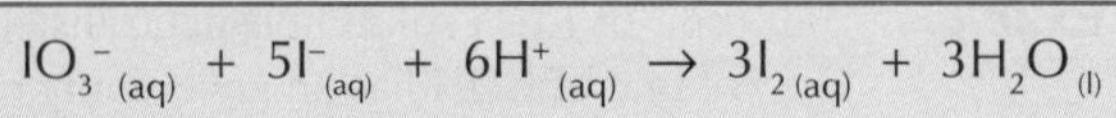

STAGE 2: Find out how many moles of iodine have been produced.

You do this by **titrating** the resulting solution with **sodium thiosulfate**.
(You need to know the concentration of the sodium thiosulfate solution.)

The iodine in the solution reacts with **thiosulfate ions** like this:

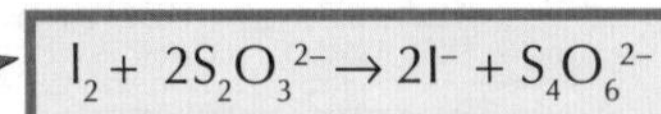

Titration of Iodine with Sodium Thiosulfate

1) Put all the solution produced in Stage 1 in a flask.
2) From the burette, add sodium thiosulfate solution to the solution in the flask.
3) It's dead hard to see the end point, so when the iodine colour fades to pale yellow, add 2 cm³ of starch solution (to detect the presence of iodine). The solution in the conical flask will go dark blue, showing there's still some iodine there.
4) Add sodium thiosulfate one drop at a time until the blue colour disappears.
5) When this happens, it means all the iodine has just been reacted.
6) Now you can calculate the number of moles of iodine in the solution.

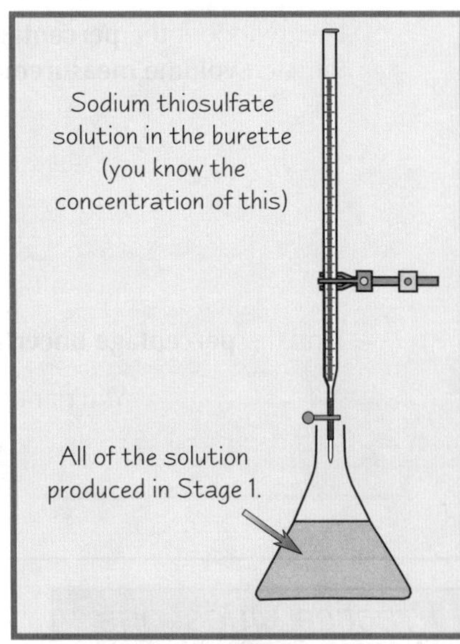

Here's how you'd do the titration calculation to find the **number of moles of iodine** produced in Stage 1.

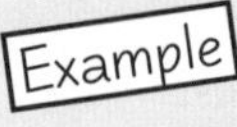

The iodine in the solution produced in Stage 1 reacted fully with 11.1 cm³ of 0.12 mol dm⁻³ thiosulfate solution.

$$I_2 + 2S_2O_3^{2-} \rightarrow 2I^- + S_4O_6^{2-}$$

11.1 cm³
0.12 mol dm⁻³ (under $2S_2O_3^{2-}$)

Number of moles of thiosulfate $= \dfrac{\text{concentration} \times \text{volume (cm}^3)}{1000} = \dfrac{0.12 \times 11.1}{1000} = \mathbf{1.332 \times 10^{-3}\ moles}$

1 mole of iodine reacts with **2 moles** of thiosulfate,

So number of **moles of iodine** in the solution $= 1.332 \times 10^{-3} \div 2 = \mathbf{6.66 \times 10^{-4}\ moles}$

STAGE 3: Calculate the concentration of the oxidising agent.

1) Now you look back at your original equation: $IO_3^-{}_{(aq)} + 5I^-{}_{(aq)} + 6H^+{}_{(aq)} \rightarrow 3I_{2\,(aq)} + 3H_2O_{(l)}$
2) The equation shows that **one mole** of iodate(V) ions produces **three moles** of iodine.
25 cm³ of potassium iodate(V) solution produced **6.66×10^{-4} moles of iodine**.
So there must have been **$6.66 \times 10^{-4} \div 3 = 2.22 \times 10^{-4}$ moles of iodate(V) ions**.
There would be the same number of moles of potassium iodate(V) in the solution. So now it's straightforward to find the **concentration** of the potassium iodate(V) solution, which is what you're after:

$$\text{number of moles} = \frac{\text{concentration} \times \text{volume (cm}^3)}{1000} \Rightarrow 2.22 \times 10^{-4} = \frac{\text{concentration} \times 25}{1000}$$

$\Rightarrow$ **concentration of potassium iodiate(V) solution = 0.00888 mol dm⁻³**

Iodine-Sodium Thiosulfate Titration

You Have to Be Able to **Evaluate** the Titration Procedure

Titrations like the one on the previous page can give very accurate results, but there are a few ways things could go pear-shaped:

1) Using contaminated apparatus could make your results inaccurate — so make sure the burette is very **clean**, and **rinse** it out with sodium thiosulfate before you start (because traces of water will dilute the solution).
2) It's important to **read the burette correctly** (from the bottom of the meniscus, with your eyes level with the liquid).
3) To reduce the effect of random errors, **repeat** the experiment and take an average.
4) But remember to **wash** the flask between experiments or use a new, clean one.

Choppy seas made it difficult for Captain Blackbird to read the burette accurately.

This particular experiment can also have some specific problems:

- The solutions you're using will react very slowly with the air, so they should be made up as freshly as possible.
- If you add the **starch** solution **too soon** during the titration, the iodine will 'stick' to the starch and won't react as expected with the thiosulfate, making the result unreliable. Only add the starch when the solution is **pale yellow**.

Practice Questions

Q1 What is added during an iodine-sodium thiosulfate titration to make the end point easier to see?

Q2 How can an iodine-sodium thiosulfate titration help you to work out the concentration of an oxidising agent?

Q3 How many moles of thiosulfate ions react with one mole of iodine molecules?

Q4 Describe the colour change at the end point of the titration.

Exam Question

Q1 10 cm^3 of potassium iodate(V) solution was reacted with excess acidified potassium iodide solution. All of the resulting solution was titrated with 0.15 mol dm^{-3} sodium thiosulfate solution. It fully reacted with 24.0 cm^3 of the sodium thiosulfate solution.

a) Write an equation showing how iodine is formed in the reaction between iodate(V) ions and iodide ions in acidic solution. [2 marks]

b) How many moles of thiosulfate ions were there in 24.0 cm^3 of the sodium thiosulfate solution? [1 mark]

c) In the titration, iodine reacted with sodium thiosulfate according to this equation:

$$I_{2(aq)} + 2Na_2S_2O_{3(aq)} \rightarrow 2NaI_{(aq)} + Na_2S_4O_{6(aq)}$$

Calculate the number of moles of iodine that reacted with the sodium thiosulfate solution. [1 mark]

d) How many moles of iodate(V) ions produce 1 mole of iodine from potassium iodide? [1 mark]

e) What was the concentration of the potassium iodate(V) solution? [2 marks]

Two vowels went out for dinner — they had an iodate...

This might seem like quite a faff — you do a redox reaction to release iodine, titrate the iodine solution, do a sum to find the iodine concentration, write an equation, then do another sum to work out the concentration of something else. The thing is though, it does work, and you do have to know how. If you're rusty on the calculations, look back to p85.

Reaction Rates

The rate of a reaction is just how quickly it happens. Lots of things can make reactions go faster or more slowly.

Particles Must Collide to React

1) Particles in liquids and gases are **always moving** and **colliding** with **each other**. They **don't** react every time though — only when the **conditions** are right. A reaction **won't** take place between two particles **unless** —

   - They collide in the **right direction**. They need to be **facing** each other the right way.
   - They collide with at least a certain **minimum** amount of kinetic (movement) **energy**.

   This stuff's called **Collision Theory**.
2) The **minimum amount of kinetic energy** particles need to react is called the **activation energy**. The particles need this much energy to **break the bonds** to start the reaction.
3) Reactions with **low activation energies** happen **pretty easily**. But reactions with **high activation energies** don't. You need to give the particles extra energy by **heating** them.

To make this a bit clearer, here's an **enthalpy profile diagram**.

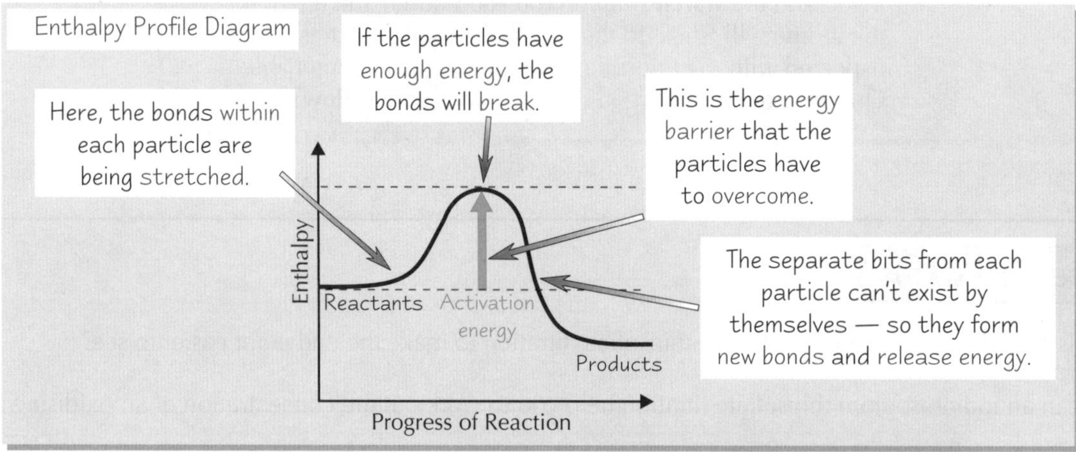

Molecules in a Gas Don't all have the Same Amount of Energy

Imagine looking down on Oxford Street when it's teeming with people. You'll see some people ambling along **slowly**, some hurrying **quickly**, but most of them will be walking with a **moderate speed**. It's the same with the **molecules** in a **gas**. Some **don't have much kinetic energy** and move **slowly**. Others have **loads of kinetic energy** and **whizz** along. But most molecules are somewhere **in between**.

If you plot a **graph** of the **numbers of molecules** in a **gas** with different **kinetic energies** you get a **Maxwell-Boltzmann distribution**. It looks like this:

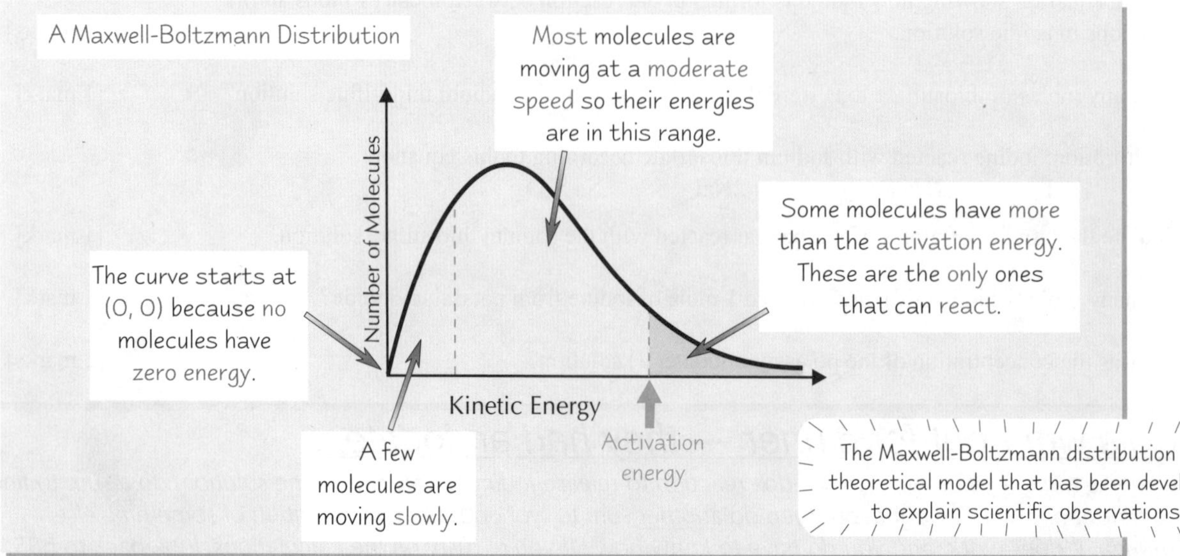

The Maxwell-Boltzmann distribution is a theoretical model that has been developed to explain scientific observations.

Reaction Rates

Increasing the Temperature or Concentration Makes Reactions Faster

1) If you increase the **temperature**, the molecules will on average have more **kinetic energy** and will move **faster**.
2) So a **greater proportion** of molecules will have energies greater than the **activation energy** (see the red bit on the diagram) and be able to **react**. This changes the **shape** of the **Maxwell-Boltzmann distribution curve** — pushing it over to the **right**.
3) And because the molecules are moving **faster**, they'll **collide more often** — **another reason** why higher temperatures make a reaction faster... and why **small temperature increases** can lead to **large increases in reaction rate**.

Higher temperature

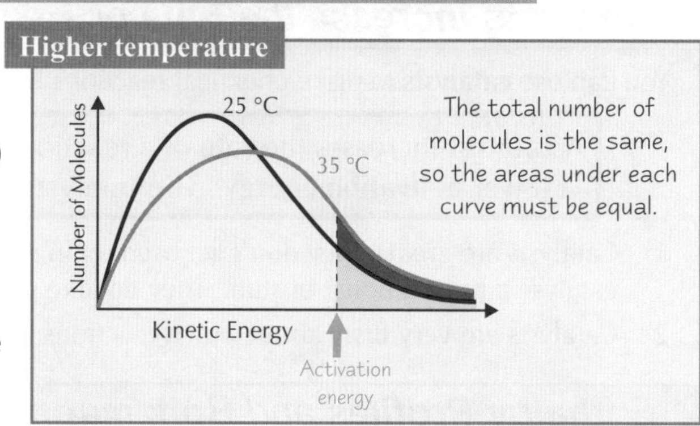

The total number of molecules is the same, so the areas under each curve must be equal.

Concentration, Surface Area and Catalysts Affect the Reaction Rate Too

Increasing Concentration (Or Pressure) Speeds Up Reactions

Increasing the **concentration** of reactants in a **solution** (or the **pressure** of a **gas**) means the particles are **closer together** on average. If they're closer, they'll **collide more often**. **More collisions** mean **more chances** to react.

Increasing Surface Area Speeds Up Reactions

If one reactant is in a **big lump** then most of the particles won't collide with other reactants.
You need to **crush** these lumps so that more of the particles can come in **contact** with the other **reactants**.
A **smaller particle size** means a **larger surface area**. This leads to a **speedier** reaction.

Catalysts Can Speed Up Reactions

Catalysts are really useful. They **lower the activation energy** by providing a **different way** for the bonds to be broken and remade. If the activation energy's **lower**, more particles will have **enough energy** to react. There's heaps of information about catalysts on **page 92**.

Practice Questions

Q1 Explain the term 'activation energy'.

Q2 Explain how and why the rate of a reaction is affected by concentration and by surface area.

Exam Questions

Q1 Nitrogen monoxide (NO) and ozone (O_3) sometimes react to produce nitrogen dioxide (NO_2) and oxygen (O_2). A collision between the two molecules does not always lead to a reaction. Explain why. [2 marks]

Q2 Use the collision theory to explain why the reaction between a solid and a liquid is generally faster than that between two solids. [2 marks]

Q3 On the right are three Maxwell-Boltzmann distribution curves.

a) Which one of the curves X and Y shows the Maxwell-Boltzman distribution curve at 15 °C ? [1 mark]

b) Use the Maxwell-Boltzman distribution curve to explain why the reaction rate may be much lower at 15 °C. [2 marks]

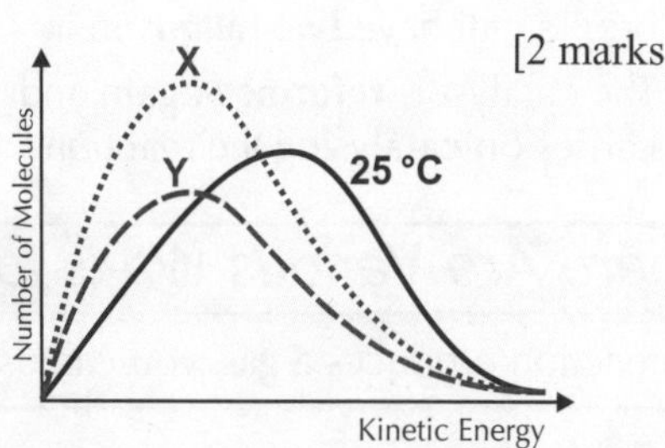

Reaction Rates — cheaper than water rates

This page isn't too hard to learn — no equations, no formulas... what more could you want? The only tricky thing might be the Maxwell-Boltzmann thingymajiggle. Remember, particles don't react every time they collide — only if they have enough energy, and are at the right angle. The more often they collide and the more energy they have, the faster the reaction is.

Catalysts and Reaction Rate Experiments

Catalysts were tantalisingly mentioned on the last page — here's the full story...

Catalysts Increase the Rate of Reactions

You can use **catalysts** to make chemical reactions happen **faster**. Learn this definition:

> A **catalyst** increases the **rate** of a reaction by providing an **alternative reaction pathway** with a **lower activation energy**. The catalyst is **chemically unchanged** at the end of the reaction.

1) Catalysts are **great**. They **don't** get used up in reactions, so you only need a **tiny bit** of catalyst to catalyse a **huge** amount of stuff. They **do** take part in reactions, but they're **remade** at the end.
2) Catalysts are **very fussy** about which reactions they catalyse. Many will usually **only** work on a single reaction.

Enthalpy Profiles and Boltzmann Distributions Show Why Catalysts Work

If you look at an **enthalpy profile** together with a **Maxwell-Boltzmann Distribution**, you can see **why** catalysts work.

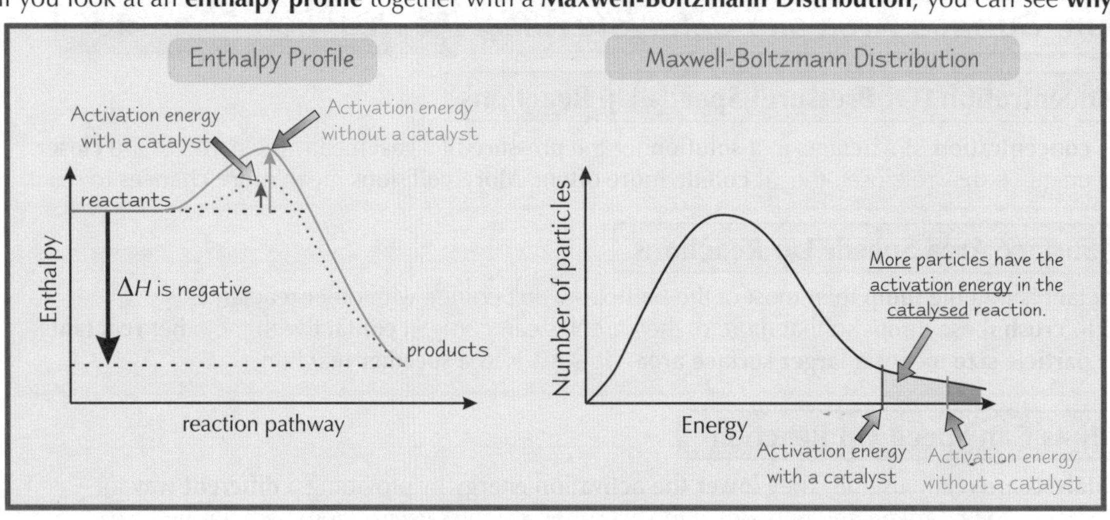

Mrs Watson tried everything to lower the camel's activation energy.

Homogeneous Catalysts Work by Forming Intermediates

1) A homogeneous catalyst is in the **same state** as the reactants. So if the reactants are **gases**, the catalyst must be a **gas** too. And if the reactants are **aqueous** (dissolved in water), the catalyst has to be **aqueous** as well.
2) A homogeneous catalyst speeds up reactions by forming one or more **intermediate compounds** with the reactants. The products are then formed from the intermediate compounds.
3) The activation energy needed to form the **intermediates** (and to form the products from the intermediates) is **lower** than that needed to make the products directly from the reactants.
4) If a reaction is speeded up by a **homogeneous catalyst**, its enthalpy profile will have **two humps** in it.
5) The catalyst is **reformed** again and carries on **catalysing** the reaction.

The Enthalpy Profile of a Homogeneously Catalysed Reaction.
(What a hideous mouthful. But it had to be said.)

E′ = the activation energy of the **first** step in the catalysed reaction.

E″ = the activation energy of the **second** step in the catalysed reaction.

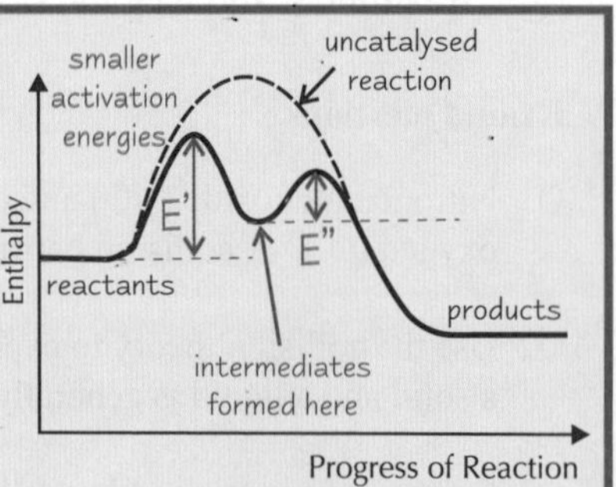

There Are Various Ways to Monitor Reaction Rate...

If a reaction produces a **gas**, you can use a **gas syringe** to record the volume of gas evolved every 10 s, say.

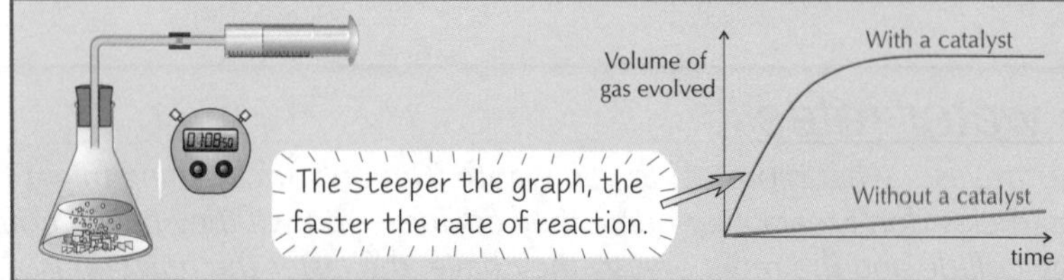

For example, this method is great for investigating the effect of the catalyst **manganese(IV) oxide** on the **decomposition of hydrogen peroxide**.

hydrogen peroxide → oxygen gas + water

O_2 is produced <u>more quickly</u> when the catalyst is added.

Catalysts and Reaction Rate Experiments

...Depending On What the **Products** of the Reaction Are

1) Another way to monitor the rate of reaction when a gas is produced is to stand the reactant vessel on a **balance** — the **mass will decrease** as gas is evolved. (This would only be suitable for **non-toxic** gases of course.)
2) Not every reaction produces a gas, though. Some produce a **precipitate** that **clouds** a solution. You can monitor this type of reaction by measuring how quickly a marker becomes invisible through the cloudiness.

The reaction between **sodium thiosulfate solution and hydrochloric acid** makes a yellow **precipitate** of **sulfur**. So you can monitor the rate of this reaction using this method.
For instance, you can investigate the effect of concentration by repeating the experiment using a **different concentration** of sodium thiosulfate solution each time. You need to keep everything else the same. You then **time** how long the mark takes to disappear each time.

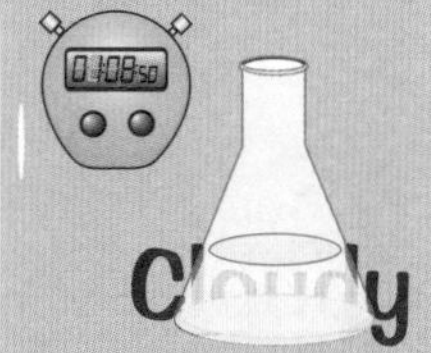

Temperature can also be investigated using this method. You warm the acid and sodium thiosulfate separately to a certain temperature, then mix them together. The **faster** the mark disappears, the **faster** the reaction.

One problem with this method is that the result's **subjective** — **different people** might not agree over the **exact** point when the mark 'disappears'. And it only works if the reactants you **started with** were transparent.

Practice Questions

Q1 Explain what a catalyst is.

Q2 Draw an enthalpy profile diagram and a Maxwell-Boltzmann distribution diagram to show how a catalyst works.

Q3 What is an 'intermediate'?

Q4 Sketch an energy profile diagram of a homogeneously catalysed reaction.

Exam Questions

Q1 Sulfuric acid is manufactured by the contact process. In one of the stages, sulfur dioxide, is converted into sulfur trioxide. A vanadium(V) oxide catalyst is used.

$$2SO_{2(g)} + O_{2(g)} \overset{V_2O_{5(s)}}{\rightleftharpoons} 2SO_{3(g)} \qquad \Delta H = -197 \text{ kJ mol}^{-1}$$

a) Draw and label an enthalpy profile diagram for the catalysed reaction. Label the activation energy. [3 marks]

b) On your diagram from part a), draw a profile for the uncatalysed reaction. [1 mark]

c) Explain what a catalysts does. [2 marks]

Q2 The decomposition of hydrogen peroxide, H_2O_2, into water and oxygen is catalysed by manganese(IV) oxide, MnO_2.

a) Write an equation for the reaction. [2 marks]

b) Sketch a Maxwell-Boltzmann distribution for the reaction.
Mark on the activation energy for the catalysed and uncatalysed process. [3 marks]

c) Referring to your diagram from part b), explain how manganese(IV) oxide acts as a catalyst. [3 marks]

I'm a catalyst — I like to speed up arguments without getting too involved...

Whatever you do, do not confuse the Maxwell-Boltzmann diagram for catalysts with the one for a temperature change. Catalysts lower the activation energy without changing the shape of the curve. BUT, the shape of the curve does change with temperature. Get these mixed up and you'll be the laughing stock of the Examiners' tea room.

Chemical Equilibria

There's a lot of to-ing and fro-ing on this page. Mind your head doesn't start spinning.

Reversible Reactions Can Reach *Dynamic Equilibrium*

1) Lots of chemical reactions are **reversible** — they go **both ways**. To show a reaction's reversible, you stick in a $\rightleftharpoons$. Here's an example:

$$H_{2(g)} + I_{2(g)} \rightleftharpoons 2HI_{(g)}$$

This reaction can go in **either direction** —

forwards $H_{2(g)} + I_{2(g)} \rightarrow 2HI_{(g)}$ ....or **backwards** $2HI_{(g)} \rightarrow H_{2(g)} + I_{2(g)}$.

2) As the **reactants** get used up, the **forward** reaction **slows down** — and as more **product** is formed, the **reverse** reaction **speeds up**.
3) After a while, the forward reaction will be going at exactly the **same rate** as the backward reaction. The amounts of reactants and products **won't be changing** any more, so it'll seem like **nothing's happening**. It's a bit like you're **digging a hole**, while someone else is **filling it in** at exactly the **same speed**. This is called a **dynamic equilibrium**.
4) A **dynamic equilibrium** can only happen in a **closed system**. This just means nothing can get in or out.

You can Predict what will Happen if *Conditions are Changed*

If you **change** the **concentration**, **pressure** or **temperature** of a reversible reaction, you tend to **alter** the **position of equilibrium**. This just means you'll end up with **different amounts** of reactants and products at equilibrium.

If the position of equilibrium moves to the **left**, you'll get more **reactants**.

$$H_{2(g)} + I_{2(g)} \rightleftharpoons 2HI_{(g)}$$

If the position of equilibrium moves to the **right**, you'll get more **products**.

$$H_{2(g)} + I_{2(g)} \rightleftharpoons 2HI_{(g)}$$

This rule tells you how the **position of equilibrium** will change if a **condition changes**:

If there's a change in **concentration**, **pressure** or **temperature**, the equilibrium will move to help **counteract** the change.

So, basically, if you **raise the temperature**, the position of equilibrium will shift to try to **cool things down**. And, if you **raise the pressure or concentration**, the position of equilibrium will shift to try to **reduce it again**.

Catalysts *Don't Affect* The Position of Equilibrium

Catalysts have **NO EFFECT** on the **position of equilibrium**. They **can't** increase **yield** — but they **do** mean equilibrium is reached **faster**.

Chemical Equilibria

Here's How to Predict **Which Way** the **Equilibrium** will **Move**

CONCENTRATION $2SO_{2(g)} + O_{2(g)} \rightleftharpoons 2SO_{3(g)}$

1) If you **increase** the **concentration** of a **reactant** (SO_2 or O_2), the equilibrium tries to **get rid** of the extra reactant. It does this by making **more product** (SO_3). So the equilibrium's shifted to the **right**.
2) If you **increase** the **concentration** of the **product** (SO_3), the equilibrium tries to remove the extra product. This makes the **reverse reaction** go faster. So the equilibrium shifts to the **left**.
3) **Decreasing** the concentrations has the **opposite effect**.

PRESSURE (changing this only affects **equilibria involving gases**)

1) **Increasing** the pressure shifts the equilibrium to the side with **fewer** gas molecules. This **reduces** the pressure.
2) **Decreasing** the pressure shifts the equilibrium to the side with **more** gas molecules. This **raises** the pressure again.

There are 3 moles on the left, but only 2 on the right. So, an increase in pressure shifts the equilibrium to the right. → $2SO_{2(g)} + O_{2(g)} \rightleftharpoons 2SO_{3(g)}$

TEMPERATURE

1) **Increasing** the temperature means **adding heat**.
   The equilibrium shifts in the **endothermic (positive ΔH) direction** to absorb this heat.
2) **Decreasing** the temperature **removes heat**.
   The equilibrium shifts in the **exothermic (negative ΔH) direction** to try to replace the heat.
3) If the forward reaction's **endothermic**, the reverse reaction will be **exothermic**, and vice versa.

This reaction's exothermic in the forwards direction. If you increase the temperature, the equilibrium shifts to the left to absorb the extra heat.

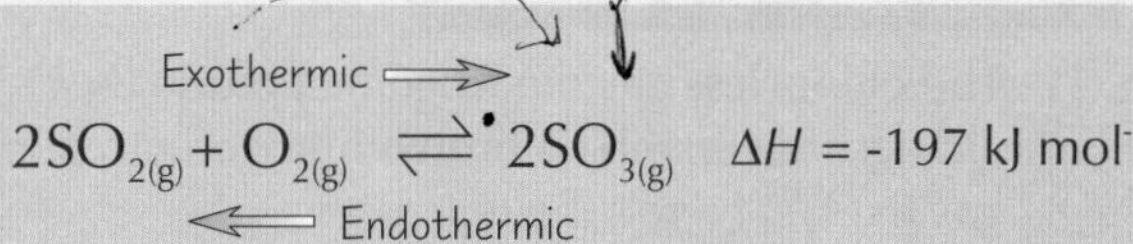

Practice Questions

Q1 Using an example, explain the terms 'reversible reaction' and 'dynamic equilibrium'.

Q2 If the equilibrium moves to the right, do you get more products or reactants?

Q3 A reaction at equilibrium is endothermic in the forward direction.
What happens to the position of equilibrium as the temperature is increased?

Exam Question

Q1 Nitrogen and oxygen gases were reacted together in a closed flask and allowed to reach equilibrium with the nitrogen monoxide formed. The forward reaction is endothermic.

$$N_{2(g)} + O_{2(g)} \rightleftharpoons 2NO_{(g)}$$

a) Explain how the following changes would affect the position of equilibrium of the above reaction:
   (i) Pressure is **increased**. [2 marks]
   (ii) Temperature is **reduced**. [2 marks]
   (iii) Nitrogen monoxide is removed. [1 mark]

b) What would be the effect of a catalyst on the composition of the equilibrium mixture? [1 mark]

Only going forward cos we can't find reverse...

Equilibria never do what you want them to do. They always ***oppose*** *you. Be sure you know what happens to an equilibrium if you change the conditions. A word about pressure — if the equation has the same number of gas moles on each side, then you can raise the pressure as high as you like and it won't make a blind bit of difference to the position of equilibrium.*

More About Equilibria

Using the principle on page 94 you can shift the position of equilibrium to make more of a product. This is really important in industry (as you'll see). But you can check it does actually work in the lab.

Simple Experiments can Show the Effect of Changes in Conditions

...And they don't even need much fancy equipment.

Changing the CONCENTRATION

1) If you put **iodine(I) chloride** (a brown liquid) in the apparatus shown and pass **chlorine gas** over it, **iodine(III) chloride** (a yellow solid) forms.
   The reaction is:

$$ICl_{(l)} + Cl_{2(g)} \rightleftharpoons ICl_{3(s)}$$

brown liquid — yellow solid

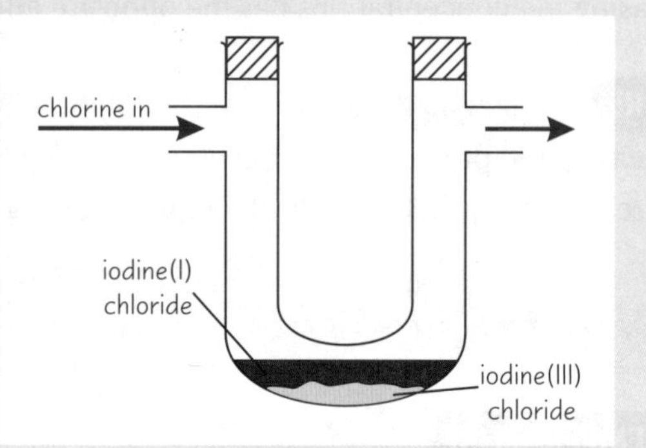

2) If you pump more chlorine into the tube, you'll **increase** its **concentration**. The equilibrium moves to the **right** — so you get more of the yellow solid.
3) If you stop the supply of chlorine and let air in, the **concentration** of chlorine **decreases**. The equilibrium moves back to the **left**, so the **yellow solid** starts turning back into the **brown liquid**.

Changing the TEMPERATURE

1) The **colourless gas** dinitrogen tetroxide (N_2O_4) changes **reversibly** to **brown** nitrogen dioxide gas (NO_2). Here's the equation:

$$N_2O_{4(g)} \rightleftharpoons 2NO_{2(g)} \quad \Delta H = +58 \text{ kJ mol}^{-1}$$

colourless gas — brown gas

It's an endothermic reaction.

2) If you stick a **sealed syringe** full of this equilibrium mixture in a beaker of **hot water**, the gas gets **darker** because more brown NO_2 forms.
3) The extra heat sends the equilibrium in the **endothermic** direction.

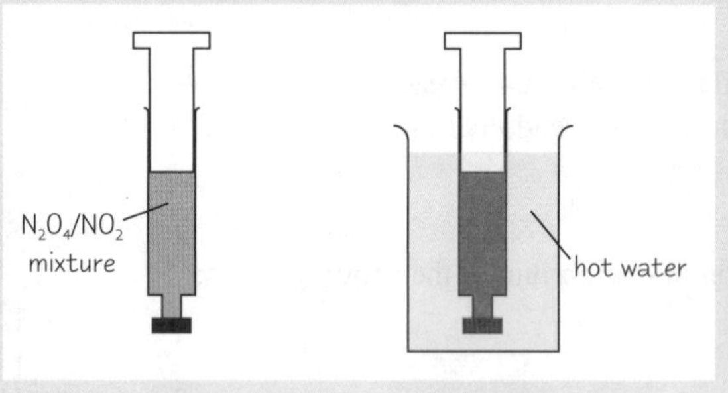

Changing the PRESSURE

1) This one also uses the syringe containing the **N_2O_4 / NO_2 mixture.**
2) If you compress the gas in the syringe, it gets **darker** at first — that's because the **concentration** has increased. But then the mixture gets **paler** as some NO_2 is converted into colourless N_2O_4.
3) The equilibrium has moved towards the side with **fewer moles** of gas to reduce the pressure again.

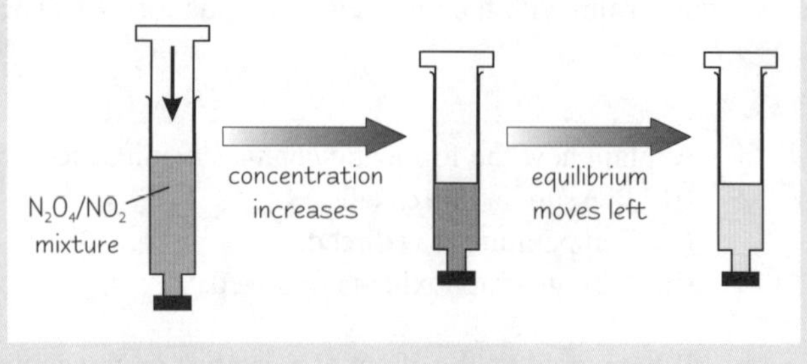

More About Equilibria

Businesses want to find the **Best Conditions** for their **Processes**

1) Predicting the effect of changing conditions is really important for **industries** that use **reversible reactions**.
2) The big bosses want to make bags of **money**. That means finding a way to make as much of the **useful products** as they can, as **cheaply** as possible.
3) So they need to pick the **best conditions** for their processes.

The **Temperature** and **Pressure** are Often **Compromises**

1) For example, look at this reaction. It's how methanol is produced from hydrogen and carbon monoxide.
You don't need to memorise the details, it's just another example of how Le Chatelier's principle is applied.

$$2H_{2(g)} + CO_{(g)} \rightleftharpoons CH_3OH_{(g)} \qquad \Delta H = -90 \text{ kJ mol}^{-1}$$

2) Because it's an **exothermic reaction**, **lower** temperatures favour the forward reaction. This means **more** hydrogen and carbon monoxide is converted to methanol — you get a better **yield**.
3) The trouble is, **lower temperatures** mean a **slower rate of reaction**. You'd be **daft** to try to get a **really high yield** if it's going to take 10 years. So 250 °C is used — a **compromise** between **maximum yield** and **a faster reaction**.
4) **Higher pressures** favour the **forward reaction**. This is because **high pressure** moves the reaction to the side with **fewer molecules of gas**. There are 3 moles of gas on the reactant side ($2H_{2(g)} + CO_{(g)}$) and only 1 mole on the product side ($CH_3OH_{(g)}$).
5) **Increasing the pressure** also increases the **rate** of reaction. Cranking up the pressure as high as you can sounds like a great idea so far. But **high pressures** are **expensive** to produce, and you need **stronger pipes**, etc. to withstand them.
6) So the **50-100 atmospheres** used is **compromise** between **maximum yield** and **expense**.
7) A **catalyst** is used to make the reaction reach equilibrium **more quickly**.

Practice Questions

Q1 How could you show using a simple experiment that changing temperature can change the position of equilibrium?

Q2 What effect does increasing the pressure have on the position of equilibrium of this reaction?

$$N_2O_{4(g)} \rightleftharpoons 2NO_{2(g)}$$

Q3 Describe how you could show that increasing the concentration of chlorine in this reaction moves the equilibrium to the right. Draw a diagram of the apparatus needed.

$$ICl_{(l)} + Cl_{2(g)} \rightleftharpoons ICl_{3(s)}$$

Q4 What effect does a catalyst have on an equilibrium reaction?

Exam Question

Q1 The manufacture of ethanol can be represented by the reaction

$$C_2H_{4(g)} + H_2O_{(g)} \rightleftharpoons C_2H_5OH_{(g)} \qquad \Delta H = -46 \text{ kJ mol}^{-1}$$

Typical conditions are 300 °C and 60-70 atmospheres.

a) Explain, in molecular terms, why a temperature lower than the one quoted is not used. [3 marks]

b) Explain why a pressure higher than the one quoted is not often used. [2 marks]

c) The gases are passed through a conversion chamber containing a catalyst of phosphoric acid, adsorbed onto the surface of silica. Describe and explain the effect of the catalyst on:
(i) the rate of production of ethanol, [3 marks]
(ii) the amount of ethanol in the equilibrium mixture. [2 marks]

It's all about money — it's what makes the world go around...

Lots of lovely stuff here folks. It just goes to show you that predicting the effect of changing the conditions isn't just something they make you learn to fill up the AS Chemistry syllabus. It has uses in real life. Everyone in the manufacturing business wants to make as much stuff as they can, as quickly as they can and as cheaply as they can. It's just a fact of life.

Alcohols

Alcohol — evil stuff, it is. I could start preaching, but I won't, because this page is enough to put you off alcohol for life...

Alcohols can be **Primary**, **Secondary** or **Tertiary**

1) The alcohol homologous series has the **general formula $C_nH_{2n+1}OH$**.
2) An alcohol is **primary**, **secondary** or **tertiary**, depending on which carbon atom the hydroxyl group **–OH** is bonded to...

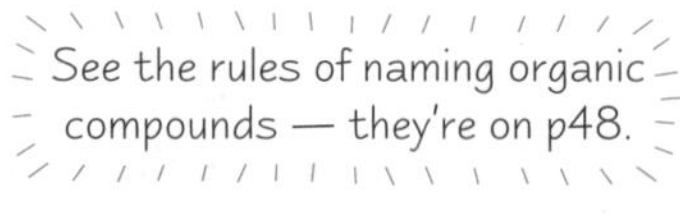
See the rules of naming organic compounds — they're on p48.

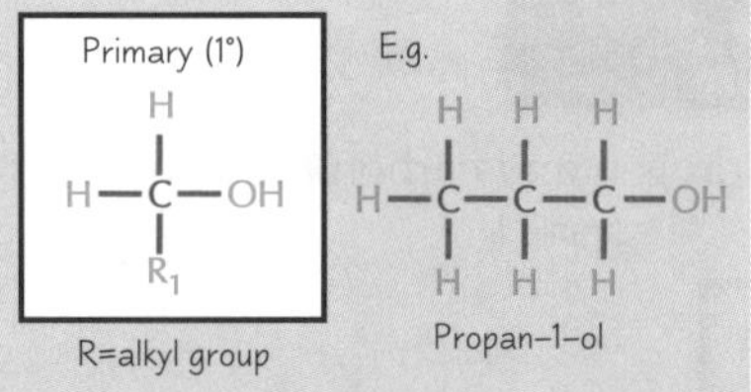

Secondary (2°)
H
R_2—C—OH
R_1

E.g.
H H H
H—C—C—C—H
H OH H
Propan–2–ol

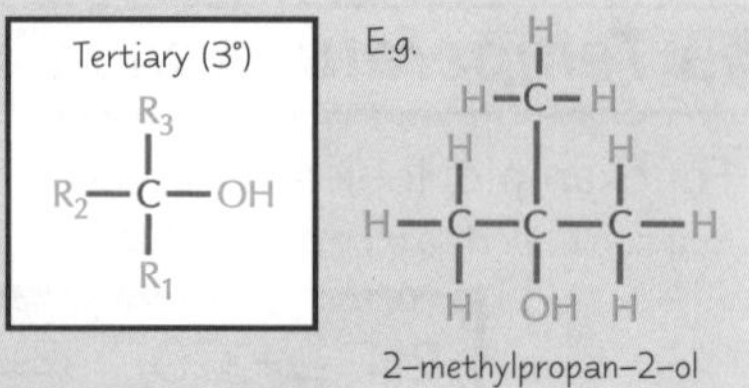

–OH can be **Swapped** for a Halogen to make a **Halogenoalkane**

1) **Alcohols** are a good starting point for making halogenoalkanes. You need to replace the alcohol's **-OH** group with a **halogen**.
2) **Tertiary** alcohols are more **reactive** than either primary or secondary alcohols, so it's easiest to start with one of these.
3) To make a **chloroalkane** you can just shake a tertiary alcohol with hydrochloric acid. This gives you an impure chloroalkane (see page 121 for how to purify it).
4) **Primary** and **secondary** alcohols react **too slowly** to be made this way. You need to use the phosphorus(III) halide method below.

Halogenoalkanes are just alkanes that have halogens in place of one or more of their hydrogens (see p103).

$$(CH_3)_2C(OH)CH_3 + HCl \longrightarrow (CH_3)_2C(Cl)CH_3 + H_2O$$

tertiary alcohol (2-methylpropan-2-ol) → haloalkane (2-chloro-2-methylpropane)

> **Bromoalkanes** and **iodoalkanes** are a bit trickier to make than chloroalkanes — HBr and HI aren't always available 'off the shelf'. Some books suggest using concentrated H_2SO_4 and a **metal halide** (e.g. KBr or KI) to produce HBr or HI 'in situ' (i.e. during the reaction process itself).
>
> The drawback is that HBr and HI are both oxidised by the H_2SO_4, so you end up with by-products (Br_2 and I_2) and a reduced yield of halogenoalkane. You get so little iodoalkane that it's better to use **phosphoric(V) acid** instead of sulfuric acid.

You Can Make Halogenoalkanes Using **Phosphorus(III) Halides** Too

1) This is the general equation: $3ROH + PX_3 \rightarrow 3RX + H_3PO_3$ X represents Cl, Br or I.
2) It's straightforward to make a **chloroalkane** by reacting an alcohol with **PCl_3**. But, **PBr_3** and **PI_3** are usually made **in situ** by refluxing (see p100) the alcohol with 'red phosphorus' and either bromine or iodine.

Chloroalkanes can Also be Made Using **Phosphorus(V) Chloride**

Here's the equation: $ROH_{(l)} + PCl_{5(l)} \rightarrow RCl_{(l)} + HCl_{(g)} + POCl_{3(l)}$

This reaction's used to **test** for alcohols...

Test for the Hydroxyl Group (–OH)

Add **phosphorus(V) chloride** to the unknown liquid.
If -OH is present, you'll get **steamy fumes** of HCl gas, which dissolve in water to form chloride ions. You can then test for chloride ions using silver nitrate (see p83).
The steamy fumes of HCl gas also turn moist **blue litmus red** (because HCl dissolves to form a strong acid).

Alcohols React with Sodium to Produce **Alkoxides**

1) **Sodium metal** reacts gently with **ethanol**, breaking the **O–H** bonds to produce ionic sodium ethoxide and hydrogen.

$$2CH_3CH_2OH + 2Na \rightarrow 2CH_3CH_2O^-Na^+ + H_2$$

2) The longer the **hydrocarbon chain** of the alcohol gets, the **less** reactive it is with sodium.

Alcohols

The Hydroxyl Group –OH can form **Hydrogen Bonds**

The **polar** –OH group on alcohols helps them to form **hydrogen bonds** (see p71), which gives them certain properties...

1) Hydrogen bonding is the **strongest** kind of intermolecular force, so it gives alcohols **high boiling points** compared to non-polar compounds, e.g. alkanes of similar sizes.

You might also hear it said that alcohols have relatively low volatility. Volatility is the tendency of something to evaporate into a gas.

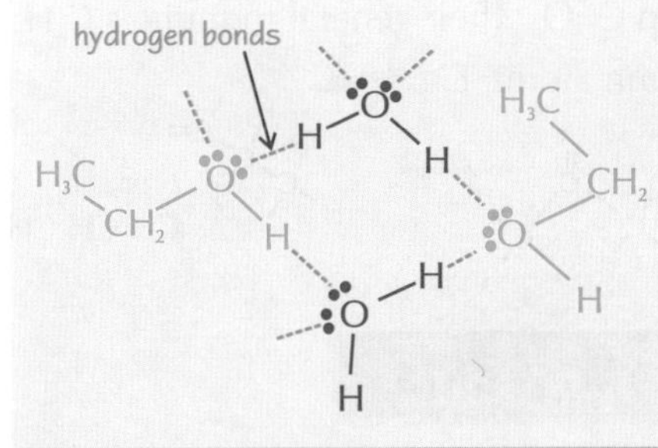

2) When you mix an alcohol with water, hydrogen bonds can also form between the **–OH** and $\mathbf{H_2O}$.
3) If it's a **small** alcohol (e.g. methanol, ethanol or propan-1-ol), hydrogen bonding lets it mix freely with water — it's **miscible** with water.
4) In **larger alcohols**, most of the molecule is a non-polar carbon chain, so there's less attraction for the polar H_2O molecules. This means that as alcohols **increase in size**, their miscibility in water **decreases.**
5) Small alcohols are also miscible in some **non-polar solvents** like cyclohexane.

Alcohols **Burn** to Produce **Carbon Dioxide** and **Water**

It doesn't take much to set ethanol alight and it burns with a **pale blue flame**. The C–C and C–H bonds are broken as the ethanol is **completely oxidised** to make carbon dioxide and water. This is a **combustion** reaction.

$$C_2H_5OH_{(l)} + 3O_{2(g)} \rightarrow 2CO_{2(g)} + 3H_2O_{(g)}$$

Practice Questions

Q1 Describe a chemical test for an alcohol.

Q2 What products are formed when a tertiary alcohol is shaken with hydrochloric acid?

Q3 Give a balanced chemical equation for the reaction between methanol and sodium, and name the organic product.

Q4 How does the boiling point of an alcohol compare to the boiling point of a similarly-sized alkane?

Exam Questions

Q1 a) Draw and name a primary alcohol, a secondary alcohol and a tertiary alcohol, each with the formula $C_5H_{12}O$. [6 marks]

b) Describe how ethanol could be converted into:
(i) bromoethane [2 marks]
(ii) sodium ethoxide [1 mark]

Q2 a) Write a balanced equation for the conversion of propan-2-ol into 2-chloropropane using phosphorus(V) chloride. [2 marks]

b) Give one safety precaution that should be taken when carrying out this reaction. [1 mark]

Q3 a) Write a balanced equation for the reaction between butan-1-ol and sodium. [2 marks]

b) Name the organic product of the reaction. [1 marks]

c) Describe any differences you would expect to observe between the reaction in a) and the reaction of methanol with sodium, and explain why the differences occur. [2 marks]

Alcohol's like education — primary, secondary, tertiary...

Back in year 8 or something, you probably had PSHE lessons about alcohol — in those days, 'learning about alcohol' meant arranging cut-out drinks in order of alcoholic strength. But now you need to know how to test for the OH group, the three different types of alcohol, how alcohols react with sodium and how to use them to make halogenoalkanes. Hmmph.

Oxidation of Alcohols

Another page of alcohol reactions. Probably not what you wanted for Christmas...

How Much an Alcohol can be **Oxidised** Depends on its **Structure**

You can use the **oxidising agent acidified potassium dichromate(VI)** to **mildly** oxidise alcohols.

- **Primary** alcohols are oxidised to **aldehydes** and then to **carboxylic acids.**
- **Secondary** alcohols are oxidised to **ketones** only.
- **Tertiary** alcohols won't be oxidised.

The orange dichromate(VI) ion is reduced to the green chromium(III) ion, Cr^{3+}.

Aldehydes and **ketones** are **carbonyl** compounds — they have the functional group C=O. Their general formula is $C_nH_{2n}O$.

1) **Aldehydes** have a **hydrogen** and **one alkyl group** attached to the carbonyl carbon atom. E.g.
2) **Ketones** have **two alkyl groups** attached to the carbonyl carbon atom.

propanone CH_3COCH_3

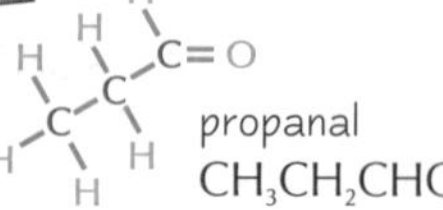

propanal CH_3CH_2CHO

Primary Alcohols will Oxidise to **Aldehydes** and **Carboxylic Acids**

$$R{-}CH_2{-}OH + [O] \longrightarrow R{-}CHO + H_2O \quad + [O] \xrightarrow{\text{reflux}} R{-}COOH$$

primary alcohol — aldehyde — carboxylic acid

[O] = oxidising agent

You can control how **far** the alcohol is oxidised by controlling the **reaction conditions**:

Oxidising Primary Alcohols — Making an Aldehyde

1) Gently heating ethanol with potassium dichromate(VI) solution and sulfuric acid in a test tube should produce "apple" smelling **ethanal** (an aldehyde). However, it's **really tricky** to control the amount of heat and the aldehyde is usually oxidised to form "vinegar" smelling **ethanoic acid**.
2) To get just the **aldehyde**, you need to get it out of the oxidising solution **as soon** as it's formed.

   You can do this by gently heating excess alcohol with a **controlled** amount of oxidising agent in **distillation apparatus**, so the aldehyde (which boils at a lower temperature than the alcohol) is distilled off **immediately**.
3) Distillation is used in many other processes to purify a product — you just collect the fraction that vaporises around the boiling point of the compound you're after.

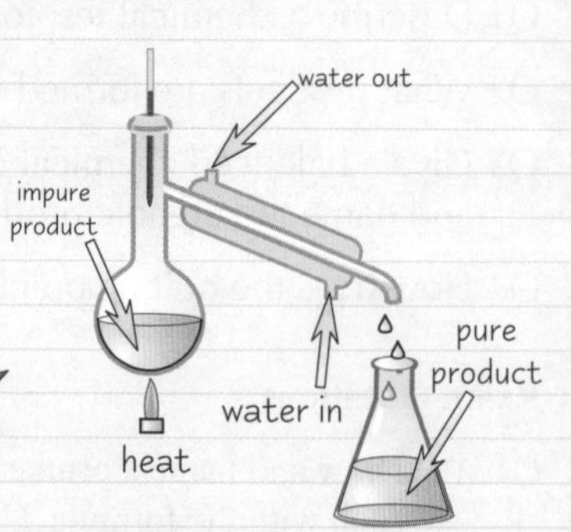

Volatile liquids (ones with a low boiling point) can be tricky to work with. If you need to heat a volatile liquid for a reaction, you can end up with most of the liquid **evaporating** before it's had chance to react.

You can get around this by fitting a **reflux condenser** — that's just a condenser fitted vertically to the reaction flask. Refluxing is a very common technique in preparing organic liquids. For example:

Oxidising Primary Alcohols — Making a Carboxylic Acid

1) To produce the **carboxylic acid**, the alcohol has to be **vigorously oxidised**. The alcohol is mixed with excess oxidising agent and heated under **reflux**.
2) Heating under reflux means you can increase the **temperature** of an organic reaction to boiling without losing **volatile** solvents, reactants or products. Any vaporised compounds are cooled, condense and drip back into the reaction mixture. Handy, hey.

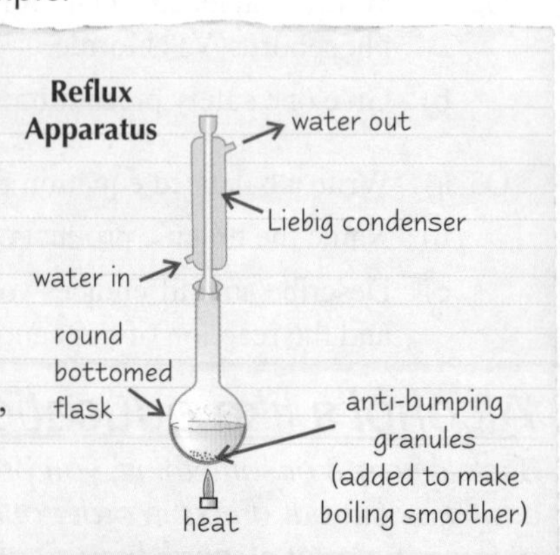

Frankie says 'reflux'

Oxidation of Alcohols

Secondary Alcohols will Oxidise to **Ketones**

$$R_1-\overset{\displaystyle H}{\underset{\displaystyle R_2}{C}}-OH + [O] \xrightarrow{\text{reflux}} \begin{matrix} R_1 \\ R_2 \end{matrix}\!\!>C=O + H_2O$$

1) Refluxing a secondary alcohol, e.g. propan-2-ol, with acidified dichromate(VI) will produce a **ketone**.
2) Ketones can't be oxidised easily, so even prolonged refluxing won't produce anything more.

Tertiary Alcohols **can't** be Oxidised Easily

Tertiary alcohols don't react with potassium dichromate(VI) at all — the solution stays orange. The only way to oxidise tertiary alcohols is by **burning** them.

Use **Oxidising Agents** to Distinguish Between **Aldehydes** and **Ketones**

Aldehydes and ketones can be distinguished using **oxidising agents** — aldehydes are easily oxidised but ketones aren't.

1) **Fehling's solution** and **Benedict's solution** are both deep blue Cu^{2+} complexes, which reduce to brick-red Cu_2O when warmed with an aldehyde, but stay blue with a ketone.
2) **Tollen's reagent** is $[Ag(NH_3)_2]^+$ — it's reduced to **silver** when warmed with an aldehyde, but not with a ketone. The silver will coat the inside of the apparatus to form a **silver mirror**.

Practice Questions

Q1 What's the difference between an aldehyde and a ketone?

Q2 What will acidified potassium dichromate(VI) oxidise secondary alcohols to?

Q3 What is the colour change when potassium dichromate(VI) is reduced?

Q4 Why are anti-bumping granules used in distillation and reflux?

Exam Question

Q1 A student wanted to produce the aldehyde propanal from propanol, and set up a reflux apparatus using acidified potassium dichromate(VI) as the oxidising agent.

a) Draw a labelled diagram of a reflux apparatus. Explain why reflux apparatus is arranged in this way. [3 marks]

b) The student tested his product and found that he had not produced propanal.

i) Describe a test for an aldehyde. [2 marks]

ii) What is the student's product? [1 mark]

iii) Write equations to show the two-stage reaction. You may use [O] to represent the oxidising agent. [2 marks]

iv) What technique should the student have used and why? [2 marks]

c) The student also tried to oxidise 2-methylpropan-2-ol, unsuccessfully.

i) Draw the full structural formula for 2-methylpropan-2-ol. [1 mark]

ii) Why is it not possible to oxidise 2-methylpropan-2-ol with an oxidising agent? [1 mark]

I.... I .... just can't do it, R2...

Don't give up now. Only as a fully-trained Chemistry Jedi, with the force as your ally, can you take on the Examiner. If you quit now, if you choose the easy path as Wader did, all the marks you've fought for will be lost. Be strong. Don't give in to hate — that leads to the dark side...

Halogenoalkanes

Don't worry if you see halogenoalkanes called haloalkanes. It's a government conspiracy to confuse you.

Halogenoalkanes are **Alkanes** with **Halogen** Atoms

Hopefully, you remember **halogenoalkanes** from 4 pages back. As the name suggests, a halogenoalkane is just an alkane with at least one **halogen atom** in place of a hydrogen atom.

E.g.

trichloromethane — 2-iodopropane — 2-bromo-2-chloro-1, 1, 1-trifluoroethane

Halogenoalkanes can be **Primary**, **Secondary** or **Tertiary**

Halogenoalkanes with just **one halogen atom** can be **primary**, **secondary** or **tertiary** halogenoalkanes.

On the **carbon** with the **halogen** attached:

1) A **primary** halogenoalkane has **two hydrogen atoms** and just **one alkyl group**.
2) A **secondary** halogenoalkane has **just one hydrogen atom** and **two alkyl groups**.
3) A **tertiary** halogenoalkane has **no hydrogen atoms** and **three alkyl groups**.

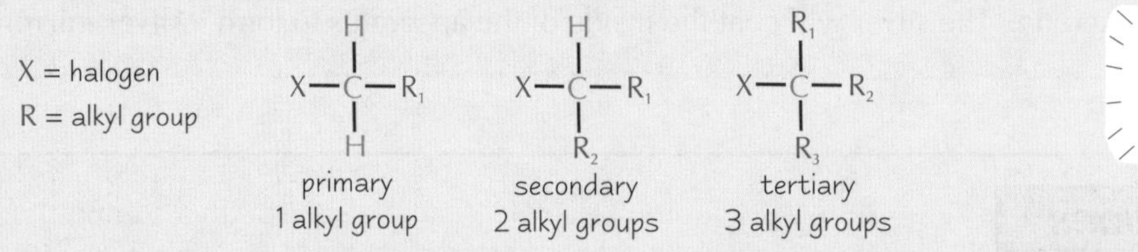

This is just the same as for alcohols.

Primary, Secondary and Tertiary Halogenoalkanes Have **Different Reactivities**

You can **compare the reactivity** of primary, secondary and tertiary halogenoalkanes by doing an experiment.

1) When you mix a **halogenoalkane** with water, it reacts to form an **alcohol**. ⟹ $R\text{–}X + H_2O \rightarrow R\text{–}OH + H^+ + X^-$
2) If you put **silver nitrate solution** in the mixture too, the silver ions react with the **halide ions** as soon as they form, giving a **silver halide precipitate** (see page 83). ⟹ $Ag^+ (aq) + X^- (aq) \rightarrow AgX (s)$
3) To compare the reactivities, set up three test tubes each containing a different halogenoalkane, ethanol (as a solvent) and silver nitrate solution (this contains the water):

The halogenoalkanes should all be isomers to make it a fair test.

50°C water bath — Start → After a few seconds → Several minutes later

A = 2-bromo-2-methylpropane (**tertiary**)
B = 2-bromobutane (**secondary**)
C = 1-bromobutane (**primary**)

4) In the tube with the **tertiary** halogenoalkane, a precipitate of silver bromide forms **immediately**.
In the tube with the **secondary** halogenoalkane, the silver bromide precipitate takes **several seconds** to form.
In the tube with the **primary** halogenoalkane, the silver bromide precipitate takes **several minutes to** form.

From the results of this experiment you can tell that the **tertiary halogenoalkane** is the most reactive, since it reacted **fastest** with the water, and the primary halogenoalkane is the least reactive.

This example uses bromoalkanes, but the order of reactivity is the same whichever halogen you use.

Halogenoalkanes

Halogenoalkanes have many **Uses**...

1) Halogenoalkanes are used to produce some rather useful **polymers** (see page 62). For example:
    - chloroethene can be used to make poly(chloroethene) — this used to be called polyvinylchloride, or PVC
    - tetrafluoroethene can be used to make poly(tetrafluoroethene), or PTFE — the non-stick coating on pans
2) Some halogenoalkanes are used as **refrigerants**, e.g. to cool the air in a fridge or air conditioning system. They're suitable for this because they're easily compressed and they don't corrode pipework.
3) Some non-flammable halogenoalkanes are used as **fire retardants** or **flame retardants**, e.g. in the plastic parts of computers or synthetic fibres in children's pyjamas.

...But their Benefits Don't Always Outweigh the **Risks**

1) As with anything in life, you have to weigh benefits against risks. Many halogenoalkanes are potentially hazardous, but the benefits (putting fires out, say) are great enough to make the risk worth taking.
2) As science progresses, the risk/benefits balance can change. This is what happened with CFCs — **chlorofluorocarbons**. CFCs are halogenoalkanes in which all the hydrogen atoms have been replaced by **chlorine** and **fluorine** atoms. They used to be used for loads of things, (e.g. refrigerants, solvents) because they're pretty **unreactive** and non-toxic.

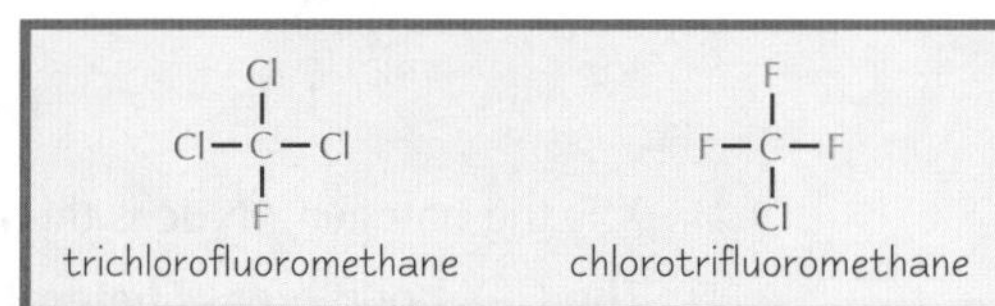

3) Then in the 1970s scientists discovered that CFCs were causing **damage** to the **ozone layer** (see page 108). The **advantages** of CFCs couldn't outweigh the **environmental problems** they were causing, so they were **banned**.
4) Many different halogenoalkanes are now used instead of CFCs. For example:
    - one widely used modern **refrigerant** (called R-410A, since you asked) is a mixture of **difluoromethane** and **pentafluoroethane**. Its big benefit is that it doesn't deplete the ozone layer. It's also non-flammable has a fairly low toxicity, but on the downside it's quite **expensive** and it's a **greenhouse gas** (see p112).
    - **hydrochlorofluorocarbons** (HCFCs) are also used in place of CFCs. They're **less damaging** to ozone because they're **less stable** and decompose lower in the atmosphere.

Practice Questions

Q1 What is a halogenoalkane?

Q2 What is a secondary halogenoalkane? Draw and name an example of one.

Q3 Put primary, secondary and tertiary halogenoalkanes in order of reactivity with water.

Q4 Give two common uses of halogenoalkanes.

Exam Questions

Q1 a) A halogenoalkane has the molecular formula C_4H_9I.
Draw and name a possible tertiary isomer of the halogenoalkane. [2 marks]

b) The halogenoalkane in part a) is mixed with water and silver nitrate solution.
Give the formula of the precipitate that forms. [1 mark]

c) Write an equation for the reaction of 1-iodobutane with water. [2 marks]

Q2 The compound 1,1,1,2-tetrafluoroethane is used as a refrigerant.

a) Draw the displayed formula of this molecule. [1 mark]

b) Give two reasons why halogenoalkanes are often used as refrigerants. [2 marks]

Hydrochlorofluorocarbon — yeah, like that's a real word...

I don't reckon there's anything too complicated here. Just learn the facts and you'll be fine. Make sure you know the difference between a primary, a secondary and a tertiary halogenoalkane. And some uses of halogenoalkanes. You might have to think about why various properties are important for different uses, or weigh up the risks and benefits.

Reactions of Halogenoalkanes

*If you haven't had enough of halogenoalkanes yet, there's more. If you **have** had enough — there's still more.*

*Halogenoalkanes May React by **Nucleophilic Substitution***

1) Halogens are much more **electronegative** than carbon. So, the **carbon–halogen bond** is **polar**.
2) The **δ+ carbon** doesn't have enough electrons. This means it can be attacked by a **nucleophile**. A nucleophile's an **electron-pair donor**. It donates an electron pair to somewhere without enough electrons.
3) **OH^-** and **NH_3** are examples of **nucleophiles** that react readily with haloalkanes. **Water** is a weak nucleophile.
4) A nucleophile can bond with the δ+ carbon of a halogenoalkane, and be **substituted** for the halogen. This is called **nucleophilic substitution**:

Here's what happens. It's a nice simple **one-step mechanism.**

$$-C^{\delta+}-X^{\delta-} + :Nuc \longrightarrow -C-Nuc + :X$$

See page 107 for heterolytic bond fission.

1) **X** is the halogen. **Nuc** is the **nucleophile**, which provides a **pair of electrons** for the **$C^{\delta+}$**.
2) The C–X bond breaks **heterolytically** — **both** electrons from the bond are taken by the halogen.
3) The halogen falls off as the nucleophile bonds to the carbon.

There are **three examples** of nucleophilic substitution you need to know. Read on.

*They React with **Aqueous Alkalis** to Form **Alcohols***

To make an **alcohol** from a halogenoalkane, you need to substitute **OH^-** for the halogen. An **aqueous alkali** supplies this — it's a **hydrolysis** reaction.

E.g. **bromoethane** can be **hydrolysed** (reacted with water) to produce **ethanol**. You have to use **warm aqueous sodium** or **potassium hydroxide** under reflux conditions or it won't work.

If you don't know what 'reflux' is check out the bottom of page 100.

Here's the equation:

$$CH_3CH_2Br + OH^- \xrightarrow[\text{reflux}]{OH^- / H_2O} C_2H_5OH + Br^-$$

And here's the mechanism:

$$CH_3CH_2^{\delta+}Br^{\delta-} + :OH^- \longrightarrow CH_3CH_2OH + :Br^-$$

***Water** Can Act as a **Nucleophile** Too*

1) The **water** molecule is a **weak nucleophile**, but it will eventually substitute for the halogen — it's just a much slower reaction than the one above.
2) You get an **alcohol** produced again. Check back to page 102 for the general equation. Here's what would happen with bromoethane:

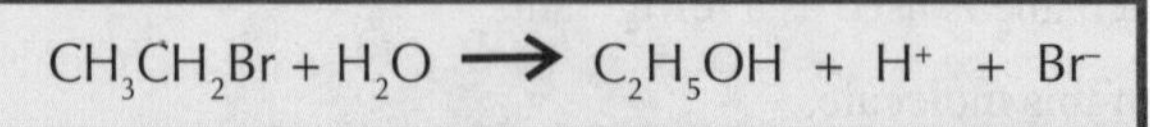

$$CH_3CH_2Br + H_2O \longrightarrow C_2H_5OH + H^+ + Br^-$$

This is another hydrolysis reaction.

And here's the mechanism:

The reaction starts in the same way — the $C^{\delta+}$ attracts a lone pair from the H_2O, and the polar C–Br bond breaks.

An intermediate forms with an oxygen that has three bonds. This is unstable, so one O–H bond breaks.

An alcohol is formed.

See page 102 for what happens if the water comes from dilute silver nitrate solution.

Reactions of Halogenoalkanes

Haloalkanes React with **Ammonia** to Form **Amines**

If you **warm** a haloalkane with excess **ethanolic** ammonia, the **ammonia** swaps places with the **halogen** — yes, it's another one of those **nucleophilic substitution reactions**.

The first step is the same as in the mechanisms on the last page, except this time the nucleophile is NH_3.

In the second step, an ammonia molecule removes a hydrogen from the NH_3 group to form an ammonium ion (NH_4^+).

The ammonium ion can react with the bromine ion to form ammonium bromide. So the overall reaction is this:

$$CH_3CH_2Br + 2NH_3 \xrightarrow[\text{ethanol}]{\text{reflux}} CH_3CH_2NH_2 + NH_4Br$$

Halogenoalkanes also Undergo **Elimination Reactions**

You know what happens when a halogenoalkane reacts with an aqueous alkali (yes you do — it's on the opposite page). But nucleophilic substitution isn't the only game in town. Swap 'aqueous' for '**alcoholic**', and things are different.

1) If you react a halogenoalkane with a warm alkali **dissolved in alcohol**, you get an **alkene**. The mixture must be **heated under reflux** or volatile stuff will be lost.
2) Here's bromoethane. Again.

$$CH_3CH_2Br + KOH \xrightarrow[\text{reflux}]{\text{ethanol}} CH_2{=}CH_2 + H_2O + KBr$$

3) This reaction has a **different mechanism** from the three previous ones. It's an **elimination** reaction — H and Br are lost and aren't replaced. (You won't be expected to draw all the details for this one, but you do have to recognise it.)

OH^- acts as a base and takes a proton, H^+, from the carbon on the left. This makes water. The left carbon now has a spare electron, so it forms a double bond with the other carbon. To form the double bond, the right carbon has to let go of the Br, which drops off as a Br^- ion.

Practice Questions

Q1 What is a nucleophile?

Q2 Sketch the mechanism, including curly arrows, for the reaction of chloroethane with warm aqueous KOH.

Q3 Write an equation for the reaction under reflux of 1-iodopropane with alcoholic potassium hydroxide.

Exam Question

Q1 Some reactions of 2-bromopropane, $CH_3CHBrCH_3$, are shown.

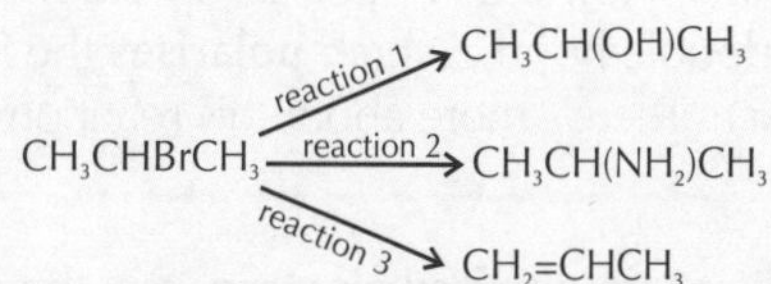

a) For each reaction, name the reagent and solvent used. [6 marks]

b) The product in reaction 1 could also be made by shaking 2-bromopropane with aqueous silver nitrate.

(i) What would be observed during the reaction? [2 marks]

(ii) Write an equation, including state symbols, to describe the reaction fully. [2 marks]

If you don't learn this — you will be eliminated...

Polar bonds get in just about every area of Chemistry. If you still think they're something to do with either bears or mints, you need to flick back to page 68 and have a good read. Make sure you learn the stuff about elimination and substitution reactions. It's always coming up in exams. Ruin the examiner's day and get it right.

Types of Reaction

This page is chock-full of really good words, like 'free radical substitution'. And 'heterolytic fission'. It's well worth a read.

You Can **Classify** Reactions by Reaction **Type**...

Here's a quick reminder of all the **reaction types** you've come across so far (in Unit 1 as well as Unit 2):

Addition – joining two or more molecules together to **form** a larger molecule.
Polymerisation – joining together lots of simple molecules to **form a giant molecule.**
Elimination – when a **small group of** atoms **breaks away** from a larger molecule.
Substitution – when **one species** is **replaced by another**.
Hydrolysis – splitting a molecule into two new molecules by **adding H^+ and OH^-** derived from **water**.
Oxidation – any reaction in which an atom **loses electrons.**
Reduction – any reaction in which an atom **gains electrons.**
Redox – any reaction where **electrons are transferred** between two species.

A **species** is an atom, an ion, or a molecule.

You Need to Know Some **Mechanisms** Too

1) Some reaction types can happen by more than one **mechanism**. Take addition, for example — you can get **nucleophilic** addition, **electrophilic** addition and **free radical** addition.
2) You're expected to **remember** the mechanisms for some particular reactions, including a couple from Unit 1:
   - **free radical substitution** of chlorine in alkanes, to make **chloroalkanes** — see p51.
   - **electrophilic addition** of bromine and hydrogen bromide to alkenes, to make **bromoalkanes** — see p58.

Classifying Reagents Helps to Predict What Reactions Will Happen

Knowing the **type of reagent** that you have helps you **predict** which chemicals will react together and what products you're likely to end up with.

1) **Nucleophiles** are **electron pair donors**. Because they're **electron rich**, they're **attracted** to places that are electron poor. So they like to react with **positive** ions. Molecules with **polar bonds** are often attacked by nucleophiles too, as they have **δ+** areas.

   Nucleophiles are attracted to the $C^{\delta+}$ atom in a **polar carbon-halogen bond**. The carbon-halogen bond breaks and the nucleophile takes the halogen's place — and that's **nucleophilic substitution** (see page 104).

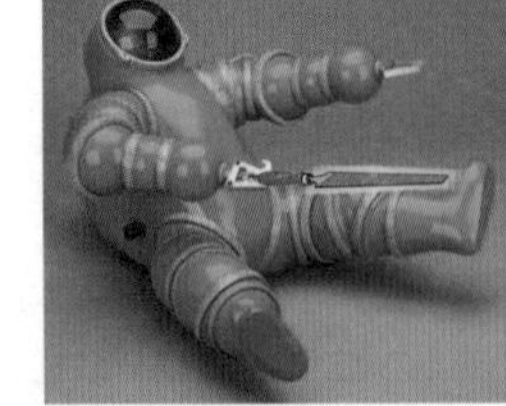
Frank put safety first when he tested his nuclear file...

2) **Electrophiles** are **electron pair acceptors**. Because they're **electron poor**, they're **attracted** to places that are electron rich. This means that they like to react with **negative** atoms and ions — and the **electron-rich** area around a **C=C bond**.

   Alkene molecules undergo electrophilic addition. In a molecule with a polar bond, like HBr, the $H^{\delta+}$ acts as an **electrophile** and is strongly attracted to the C=C double bond, (which **polarises** the H–Br bond even more, until it finally breaks). There's more about this reaction on pages 56 and 58.

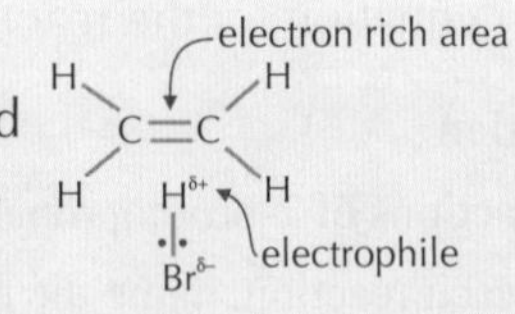

3) **Free radicals** have an **unpaired electron**, e.g. the chlorine atoms produced when UV light splits a Cl_2 molecule. Because they have unpaired electrons, they're very, very **reactive**. Unlike electrophiles and nucleophiles, they'll react with anything, positive, negative or neutral.

   UV

   $Cl{-}Cl \rightarrow 2Cl\cdot$

   Because a free radical will react with anything in sight, you'll probably end up with a mixture of products. So free radical reactions aren't much use if you're after a pure product.

   Free radicals will even attack stable non-polar bonds, like C–C and C–H (so they're one of the few things that will react with alkanes). There's loads about the reactions of free radicals with alkanes on page 51.

Types of Reaction

There are Two Types of Bond Fission — Homolytic and Heterolytic

Breaking a covalent bond is called **bond fission**. A single covalent bond is a shared pair of electrons between two atoms. It can break in two ways:

Heterolytic Fission

1) The bond breaks 'unevenly'. **Both electrons** from the shared electron pair move to **one atom.**
2) This forms two **different** species ('hetero' means 'different'):
   - a positively charged **cation** (X^+) — an **electrophile**
   - a negatively charged **anion** (Y^-) — a **nucleophile**

$$X{\div}Y \rightarrow X^+ + Y^-$$

Y gets both the electrons, so it becomes the nucleophile and X becomes the electrophile.

Homolytic Fission

1) The bond breaks evenly. **One electron** moves to **each atom.**
2) This forms two electrically uncharged **free radicals** — both atoms now have an unpaired electron.

$$X{\div}Y \rightarrow X\bullet + Y\bullet$$

3) Because of the unpaired electron, free radicals are very reactive.

A double-headed arrow shows that a pair of electrons move. A single-headed arrow shows the movement of a single electron. Makes sense.

Practice Questions

Q1 Explain why the reaction between ethene and hydrogen bromide to produce bromoethane can be described as an electrophilic addition reaction.

Q2 Bromoethane reacts with sodium hydroxide in ethanol to give ethene. What type of reaction is this?

Q3 What does 'heterolytic fission' mean?

Exam Questions

Q1 a) Many reaction mechanisms involve radicals. Explain what is meant by a 'radical' [1 mark]

b) A reaction between methane and chlorine can be initiated by strong sunlight, and involves the homolytic fission of chlorine.
Show the mechanisms of:

(i) the initiation step in this reaction [1 mark]

(ii) the propagation step in this reaction and [1 mark]

(iii) a termination step in this reaction. [1 mark]

Q2 Bromoethane can be converted into ethylamine by a two-step procedure.

a) What type of reaction is the first step? [1 mark]

b) State the reagents and conditions for the conversion. [3 marks]

Q3 Lithium aluminium hydride is a common reducing agent in organic chemistry.
Give the name and structural formula of the products formed when the following molecules are reduced:

a) ethanal [2 marks]

b) propanone [2 marks]

Reduction is just the opposite of oxidation.

Oxidising agent SALE NOW ON — everything's reduced...

Scientists do love to classify everything, and have it neatly in order. I knew one who liked to alphabetise his socks. But that's a whole other issue. Just learn the definitions for the types of reactions and reagents — and what types of reagent undergo what types of reaction. Then you'll have this page pretty much sorted. Without having to alphabetise anything.

The Ozone Layer

The ozone layer seems to have been forgotten about lately, with all the worries about climate change. It's still there though.

The Earth has a Layer of **Ozone** at the Edge of the **Stratosphere**

1) The **ozone layer** is in a layer of the atmosphere called the **stratosphere**. It contains most of the atmosphere's **ozone molecules**, O_3.
2) The ozone layer removes the dangerous high energy **UV radiation**. This can damage the DNA in cells and cause **skin cancer**. It's the main cause of **sunburn** too.
3) Now for the hard chemistry...
   Ozone is formed when **UV radiation** from the Sun hits oxygen molecules.

If the right amount of **UV radiation** is absorbed by an oxygen molecule, the oxygen molecule splits into separate atoms or **free radicals** (see page 107). The free radicals then **combine** with other oxygen molecules to form **ozone molecules**, O_3.

$$O_2 + h\nu \rightarrow O\bullet + O\bullet \Longrightarrow O_2 + O\bullet \rightarrow O_3$$

UV radiation ($h\nu$)

The Ozone Layer is Constantly Being **Replaced**

1) UV radiation can also **reverse** the formation of ozone. $O_3 + h\nu \rightarrow O_2 + O\bullet$ — The radical produced then forms more ozone with an O_2 molecule, as shown above.
2) So, the ozone layer is continuously being **destroyed** and **replaced** as UV radiation hits the molecules. An **equilibrium** is set up, so the concentrations stay fairly constant: $O_2 + O\bullet \rightleftharpoons O_3$

Scientists Discovered that the **Ozone Layer** Was a **Bit Thin** in Places

1) In the 1970s, a team from the **British Antarctic Survey** found that the concentration of ozone over Antarctica was very low compared to previous measurements. In 1985 they measured it again and they found that it was **even lower**.
2) The decrease was so **dramatic** that they thought their measuring instruments were faulty. They got some **new instruments**, but these gave the same results. Eeeek.
3) As all good scientists do, they **published** their results so that others could check them out.
4) A **satellite** had mapped the ozone levels at about the same time. But it was programmed to treat measurements below a certain value as **errors** and to ignore them — so this evidence for the thinning of the ozone layer was **overlooked**. When the British Antarctic Survey published their findings the satellite data was re-examined and found to show the 'hole' too.
5) The ozone layer over the **Arctic** has been found to be thinning too. These 'holes' in the ozone layer are bad because they allow more harmful **UV radiation** to reach the Earth.

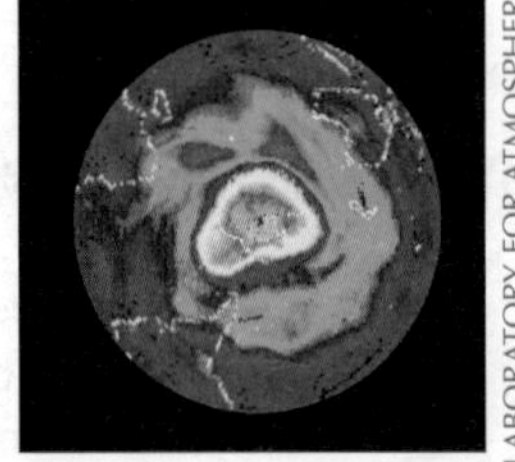

LABORATORY FOR ATMOSPHERES, NASA GODDARD SPACE FLIGHT CENTER/SCIENCE PHOTO LIBRARY

Here's a satellite map showing the 'hole' in the ozone layer over Antarctica. The 'hole' is shown by the white and pink area.

The **Ozone Layer** was being **Destroyed** by **CFCs**

It was **CFCs** (see the next page) that were breaking down the ozone layer). This is what was happening:

1) **Chlorine free radicals**, $Cl\bullet$, are formed when **CFCs** are broken down by **ultraviolet radiation**. It's a carbon-chlorine bond that's broken. E.g. $CCl_3F_{(g)} \rightarrow CCl_2F\bullet_{(g)} + Cl\bullet_{(g)}$
2) These free radicals are **catalysts**. They react with **ozone** to form an **intermediate** ($ClO\bullet$), and an oxygen molecule.

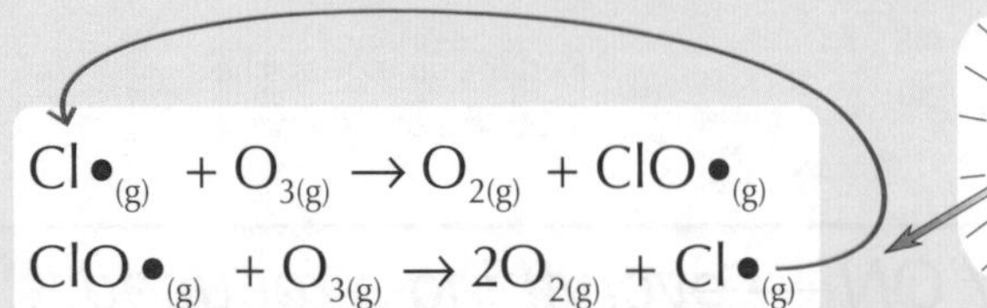

The chlorine free radical is regenerated. It goes straight on to attack another ozone molecule. It only takes one little chlorine free radical to destroy loads of ozone molecules.

3) So the **overall reaction** is... $2O_{3(g)} \rightarrow 3O_{2(g)}$ ... and $Cl\bullet$ is the catalyst.

The Ozone Layer

CFCs Have Really Great Properties — But Alternatives Had to be Found

1) **CFCs (chlorofluorocarbons)** are a group of compounds made by replacing all of the hydrogen atoms in alkanes with chlorine and fluorine. They're **halogenoalkanes** — see page 102.
2) They're **unreactive**, **non-flammable** and **non-toxic**. They were used in fire extinguishers, as propellants in aerosols, as the coolant gas in fridges and to foam plastics to make insulation and packaging materials.
3) The **Montreal Protocol** of 1989 was an **international treaty** to phase out the use of CFCs and other ozone-destroying halogenoalkanes by the year 2000. There were a few **permitted uses** such as in medical inhalers and in fire extinguishers used in submarines.
4) Scientists supported the treaty, and worked on finding **less environmentally-damaging alternatives** to CFCs (see p103).
5) The ozone holes **still** form in the spring but the **rate of decrease** of ozone is **slowing** — so things are looking up.

Nitrogen Oxides Break Ozone Down Too

Nitric oxide is sometimes called nitrogen monoxide.

1) **Nitric oxide**, **NO•**, **free radicals** destroy ozone too.
2) **NO• free radicals** come from **nitrogen oxides**, which are produced by **car and aircraft engines** and **thunderstorms**.
3) NO• free radicals affect ozone in the **same way** as chlorine radicals do. They act as catalysts:

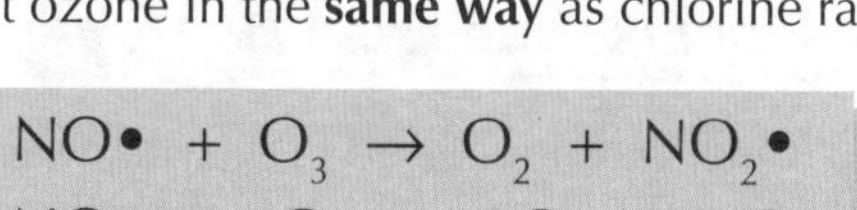

$$NO\bullet + O_3 \rightarrow O_2 + NO_2\bullet$$
$$NO_2\bullet + O_3 \rightarrow 2O_2 + NO\bullet$$

The NO• radical is regenerated — it's a catalyst.

Practice Questions

Q1 What is ozone, and where is the ozone layer?

Q2 Write out equations to show how ozone is destroyed by chlorine free radicals.

Q3 How do chlorine free radicals form?

Q4 Explain why this is an example of catalysis.

Q5 Where do nitrogen monoxide free radicals come from?

Exam Questions

Q1 The 'ozone layer' lies mostly between 15 and 30 km above the Earth's surface.

a) Explain how ozone forms in this part of the atmosphere. [3 marks]

b) What are the benefits to humans of the ozone layer? [2 marks]

c) How does the ozone layer absorb harmful radiation without being permanently destroyed? [2 marks]

Q2 CFCs were invented in 1928. They were widely used in the 20th century.

a) Give three important uses of CFCs. [3 marks]

b) What useful properties do CFCs have? [3 marks]

c) Why was the use of CFCs banned by the Montreal Protocol? [1 mark]

Q3 The high temperatures and pressures found near lightning bolts cause oxygen and nitrogen to react, forming nitrogen monoxide and nitrogen dioxide.

a) State which gas mentioned above catalyses the break down of the ozone layer. Write equations to show how this happens. [4 marks]

b) Name another source of this ozone-destroying gas. [1 mark]

A round of applause for nitric oxide — 'Molecule of the Year', 1992...

No, it really was...

How scientists found the hole in the ozone layer, repeated their experiments, then published their results is a super example of How Science Works. What's more, the evidence was used to instigate an international treaty — it's a beauty of an example of how science informs decision-making. And remember — think about any anomalous results before chucking them away.

Green Chemistry

'Green' things are big news these days — they're everywhere. So it'll be no surprise to find them in AS Chemistry too.

Chemical Industries** Could Be More **Sustainable

1) Doing something **sustainably** means you do it **without stuffing things up** for future generations. So sustainable chemistry (or 'green chemistry') means **not using up** all the Earth's **resources**, and not putting loads of **damaging** chemicals into the environment.
2) Many of the chemical processes used in industry at the moment **aren't** very sustainable. Take the **plastics** industry, for example — the raw materials used often come from non-renewable **crude oil**, and the products themselves are usually **non-biodegradable** or **hard to recycle** when we're finished with them. (See pages 62 for more details.)
3) But there are things chemists can do to try and improve things. For example, they can...

*1) Use **Renewable Raw Materials***

Loads of chemicals are traditionally made from **non-renewable** raw materials (e.g. crude oil fractions, or metal ores). But chemists can often develop **alternative compounds** (or **alternative ways** to make existing ones) involving **renewable** raw materials — e.g. some plastics are now made from **plant products** rather than oil fractions (p63).

*2) Use **Renewable** Energy Sources, or just **Use Less Energy***

1) Energy **efficiency** can be improved too. One technique used in the pharmaceutical industry is to use microwave radiation to heat the reacting mixture **directly**. (Conventional heating systems heat the **reaction vessel**, which then 'passes on' the heat to the reaction mixture — a less efficient system.)
2) Many chemical processes use a lot of **energy**. At the moment, most of that energy comes from **fossil fuels**, which will soon run out. But there are potential **alternatives**...
   - **Plant-based fuels** can be used (e.g. bioethanol — see page 114 for more information).
   - **Solar power**, **wind power**, etc.

*3) Ensure All the Chemicals Involved are as **Non-Toxic** as Possible*

1) Many common chemicals are **harmful** — either to **humans**, other **living things**, the **environment**, or all three. Where possible, it's generally a good thing to use a **safer** alternative. For example...
   - **Lead** (which can have some pretty unpleasant effects on your health) used to be used in paint, petrol and in solder for electrical components. But this meant lead got into the air from flaking paint, vehicle exhausts, and so on. Alternatives are now available — lead-free pigments are used in paints, unleaded petrol is standard now, and soldering is usually done with a mixture of tin, copper and silver.
   - Some **foams** used in fire extinguishers are very good at putting out fires, but leave hazardous products behind, including some that deplete the ozone layer. Again, alternatives are now available.
   - **Dry cleaners** used to use a solvent based on chlorinated **hydrocarbons**, but these are known to be **carcinogenic** (i.e. they cause cancer). Safer alternatives are now available (liquid 'supercritical' carbon dioxide, as you asked).
2) Sometimes **redesigning a process** means you can do without unsafe chemicals completely — e.g. instead of using harmful organic solvents, some reactions can be carried out with one of the **reactants** acting as a solvent.

*4) Make Sure that Products and Waste are **Biodegradable** or **Recyclable***

1) Chemists can also try to create **recyclable** products — a good way to conserve supplies of raw materials.
2) The amount of **waste** produced should also be kept to a **minimum**, and where possible it should be **recyclable** or **biodegradable**.
3) **Laws** can be used to encourage change. For example, when you buy a new TV, the shop now has to agree to recycle your old TV set, with the TV manufacturers paying some of the cost. This creates an incentive to design products that are easier and cheaper to recycle.

Plastics are hard to recycle. Dogs too.

Green Chemistry

Catalysts and *High Atom Economy* are Important

1) For efficiency reasons, it's good if chemical reactions have a high **atom economy** (see pages 16-17) — this **reduces waste**, and makes the best use of **resources**.
2) Scientists can also improve the **efficiency** of a process by developing new **catalysts**.
   - A new catalyst may **speed up** an **unusably slow** (but otherwise efficient) reaction enough to make it usable.
   - A new catalyst might mean you can use a **lower** (and so more energy efficient) **temperature and pressure**, but still get your products reasonably **quickly**.
3) For example, the industrial production of **ethanoic acid** (CH_3COOH) has become much more efficient over the years...

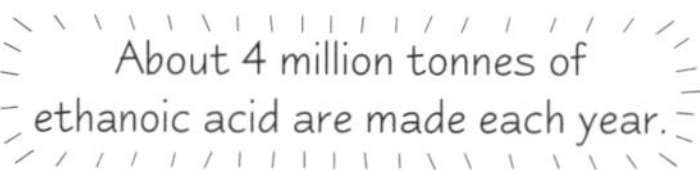

(1) Ethanoic acid was first made on an **industrial scale** by the oxidation of butane or naphtha (crude oil fractions). The reaction needed conditions of 150-200 °C, 40-50 atm pressure and a **cobalt catalyst**.

The **atom economy** of this reaction was **low** (only about 35%) because lots of other products were made too, including methanoic acid, propanone and propanoic acid. (Most of these 'side-products' are actually useful, but <u>separating</u> everything took <u>lots</u> of energy.)

(2) In 1963 the chemical company BASF developed a process using methanol and carbon monoxide:

$$CH_3OH + CO \rightarrow CH_3COOH$$

The **atom economy** of this reaction is **100%** — all the reactant molecules end up as product. So it was a much more efficient use of resources than the previous method.

But it needed a higher temperature and a much higher pressure — about 300 °C and 700 atm. Using these conditions and a **cobalt iodide catalyst** gave a yield of about 90%.

See page 15 for info about percentage yields.

(3)
1) By 1970 a different company, Monsanto, had developed a **rhodium iodide** catalyst to use with the same reaction. With this new catalyst, **less extreme conditions** were needed — 150-200 °C and 30-60 atm. These conditions meant an improved yield of about 98%.
2) The 'Monsanto process' has been the main method used to produce ethanoic acid ever since.
3) At the moment, the **methanol** that's needed is derived from crude oil. However, it could be obtained from **biomass**, including household waste — this would make the process even more sustainable.
4) Recently BP Chemicals have developed the 'Cativa™ process'. It's based on the same reaction but uses different conditions and an **iridium iodide** catalyst. This process produces **fewer by-products** than the Monsanto Process, and makes **more efficient use** of resources.

Practice Questions

Q1 List four ways in which the chemical industry can be made more sustainable.

Q2 Explain why plants are a sustainable resource.

Q3 Why is a high atom economy desirable in chemical reactions?

Exam Question

Q1 Much research is currently done on new catalysts.

a) Explain why catalysts are important in making chemical processes 'greener'. [2 marks]

b) The discovery of a new catalyst has made it possible to make ethanoic acid very efficiently by reacting methanol with carbon monoxide:

$$CH_3OH + CO \rightarrow CH_3COOH$$

Describe one way in which this reaction could be considered 'green'. [2 marks]

Like the contents of my fridge, Chemistry's going greener by the day...

It's important stuff, all this. It'll be important for your exam, obviously, but it's my bet that you'll come across this stuff long after you've taken your exam as well, which makes it doubly useful. On a different note... isn't it weird how you can sign up for an AS level in Chemistry, and only then be told that you'll be studying international politics too...

Climate Change

Now I'm sure you know this already but it's good to be sure — the greenhouse effect, global warming and climate change are all different things. They're linked (and you need to know how) — but they are not the same. Ahem.

The Greenhouse Effect Keeps Us Alive

1) The Sun emits **electromagnetic radiation**, mostly as visible light, UV radiation and infrared radiation.
2) When radiation from the Sun reaches **Earth's atmosphere**, most of the UV and infrared is **absorbed by atmospheric gases**, and some radiation is **reflected back into space** from **clouds**.
3) The energy that reaches the **Earth's surface** is mainly **visible light**, with some UV and a little infrared. Some of this radiation is reflected into space by light-coloured, shiny surfaces like ice and snow. The rest is **absorbed** by the Earth, which causes it to heat up.
4) The Earth then **radiates energy** back towards space as **infrared radiation** (heat).
5) Various gases in the troposphere (the lowest layer of the atmosphere) **absorb** some of this infrared radiation... and **re-emit** it in **all directions** — including back towards Earth, keeping us warm.
6) This is called the '**greenhouse effect**' (even though a real greenhouse doesn't actually work like this, annoyingly). Without this absorption and re-emission of heat by 'greenhouse gases', the average surface temperature on Earth would be about 30 °C cooler than it is — and we wouldn't be here.

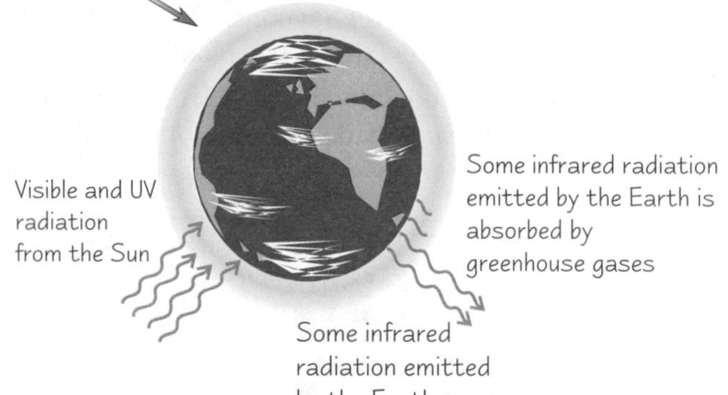

A bit more on greenhouse gases

1) The main greenhouse gases are **water vapour**, **carbon dioxide** and **methane**. They're greenhouse gases because their molecules **absorb IR radiation** to make the bonds in the molecule **vibrate more** (see page 116).
2) This extra vibrational energy is passed on to other molecules in the air by **collisions**, giving the other molecules more kinetic energy and so raising the overall temperature.
3) The greenhouse gas you hear about all the time is **carbon dioxide**, but actually water vapour makes a far greater contribution to the effect. The contribution of any particular gas depends on:
   - how much radiation one molecule of the gas absorbs
   - how much of that gas there is in the atmosphere (concentration in ppm, say)

   For example, one methane molecule traps far more heat than one carbon dioxide molecule, but there's much **less methane** in the atmosphere, so its overall contribution to the greenhouse effect is smaller.

An Enhanced Greenhouse Effect Causes Global Warming...

1) You don't see newspaper headlines about H_2O emissions — even though water vapour's responsible for a large portion of the greenhouse effect — but we're constantly being encouraged to reduce our **CO_2 emissions**.
2) That's because the concentration of water vapour in the atmosphere has stayed pretty constant for many years, whereas the concentration of carbon dioxide (and methane) has **increased** and **is still increasing**.
3) Over the last 150 years or so, the world's **human population** has shot up and we've become more **industrialised**. To supply our energy needs, we've been **burning fossil fuels** at an ever-increasing rate, releasing **tons and tons** of CO_2 into the atmosphere. We've also been **chopping down forests** which used to absorb CO_2 by photosynthesis.
4) And it's not just carbon dioxide. **Methane** levels have also risen as we've had to grow more food for our rising population. **Cows** are responsible for large amounts of methane. From both ends. (There are, it turns out, quite simple ways to reduce the problem by altering their diet.)

Vegetarians can't feel entirely smug though. Paddy fields, in which rice is grown, kick out a fair amount of methane too.

5) These **human activities** have caused a rise in greenhouse gas concentrations, which **enhances** the greenhouse effect. So now **too much heat** is being trapped and the Earth is **getting warmer** — this is **global warming**.

Climate Change

Climate Change Isn't New...

The Earth's climate has changed quite **naturally** throughout history, on various different timescales:

1) For example, regular changes in the Earth's orbit around the Sun are linked to **ice age** cycles — long cold periods (ice ages) with warmer periods (**interglacials**) in between. (We're in an interglacial period now).

2) Various changes in the **Sun**'s activity (e.g. **sunspot** cycles every 11 years) also cause warming or cooling.
3) Not all natural changes are caused by regular cycles. For example, huge **volcanic eruptions** or **meteor impacts** have thrown vast amounts of smoke or dust into the air and caused significant global cooling.

... But Anthropogenic Change Is

A lot of scientific evidence shows that global warming **is** taking place now, more quickly than in the past.

1) For example, scientists regularly sample the air in unpolluted places (like remote islands). Both average temperatures and carbon dioxide levels are going up (they both change over the seasons, of course, so it's the yearly averages that count).
2) Monitoring of **sea water** shows that the oceans have become **more acidic** as more carbon dioxide dissolves in the water (because it forms carbonic acid, H_2CO_3). So we know that the chemistry of the oceans is changing.
3) Scientists have used **mass spectrometry** to analyse the composition of **air** trapped inside the ice in polar regions, to see how the atmosphere has changed in the past (when older, deeper ice formed), and compare that with changes in recent years.
4) Putting all the evidence together, it seems to show that the Earth's average temperature has increased **dramatically** in the last 50 years, and that carbon dioxide levels have increased at the same time.
5) The **correlation** between CO_2 and temperature is pretty clear, ⟹ but there's been debate about whether rising carbon dioxide levels have **caused** the recent temperature rise. Just showing a correlation doesn't prove that one thing causes another — there has to be a plausible mechanism for how one change causes the other (in this case, the explanation is the enhanced greenhouse effect).
6) There is a consensus amongst climate scientists that the link **is** causal, and that recent warming is **anthropogenic** — **human activities** are to blame.

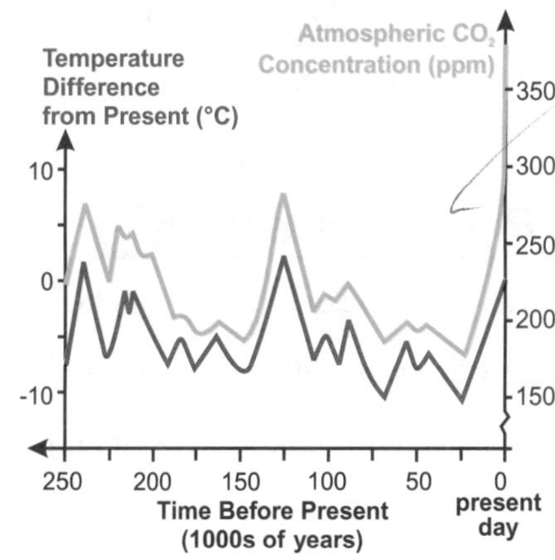

Learn What Carbon Footprint and Carbon Neutral Mean

A **carbon footprint** is the amount of greenhouse gases something causes to be released. So the more carbon dioxide an activity causes to be released, the bigger its carbon footprint.

Carbon neutral activities have **no overall carbon emission** into the atmosphere.

A couple of overturned bins make a great alternative to motor transport

1) All **products** have a **carbon footprint**. It takes **energy** to extract the raw materials, make the thing and then transport it. The energy more than likely comes from **burning fossil fuels**.
2) **Everyone** has a carbon footprint. Most of the things you do involve carbon being emitted somehow — e.g. **watching TV** and **travelling in cars or buses** involves energy from burning fossil fuels. And then everything you **buy**, including food, adds to your carbon footprint.
3) **Trees** remove CO_2 during photosynthesis. So, you can make an activity or product **carbon neutral** by planting enough trees to remove all of the CO_2 that's emitted in doing the activity or making and transporting the product.

Climate Change

Burning Most Fuels has a **Carbon Footprint**

Petrol is definitely NOT carbon neutral

Burning petrol releases CO_2 into the atmosphere that was trapped in the earth millions of years ago.

Bioethanol is carbon neutral (more or less)

Bioethanol is a possible substitute for petrol. It's produced by the **fermentation of sugar** from crops such as maize. It's thought of as being **carbon neutral**, because all the CO_2 released when the fuel is burned was removed by the crop as it grew. **BUT** — there are still carbon emissions if you consider the **whole** process. Making the fertilisers and powering agricultural machinery will probably involve burning fossil fuels.

A potential problem with using crops to make fuels is that developed countries (like us) will create a huge demand as they try and find fossil fuel alternatives. Poorer developing countries (in South America, say) will use this as a way of earning money and rush to convert their farming land to produce these 'crops for fuels', which may mean they won't grow enough food to eat.

Hydrogen gas can be pretty much carbon neutral

Hydrogen gas can either be **burned in a modified engine**, or used in a **fuel cell**. A fuel cell converts hydrogen and oxygen into **water**, and this chemical process produces **electricity** to power the vehicle. Either way, the big advantage is that **water** is the **only** waste product. Hydrogen can be extracted from **water** — you need energy to extract it though. If this energy comes from a **renewable source**, say wind or solar, it will be pretty much **carbon neutral** (but there'll be some carbon emitted when making the solar panels or wind turbines).

The trouble is, people are worried about using such a flammable gas — an airship, called the 'Hindenburg', which contained hydrogen gas exploded in 1937. But with new technology, fuel cells are very safe. Petrol is also very flammable, but people are accustomed to using it in their daily lives. On top of this, hydrogen cars cost lots more than petrol cars at the moment. Another technical problem is safely storing large quantities of hydrogen gas at refuelling stations.

Practice Questions

Q1 What type of electromagnetic radiation does the Earth emit?

Q2 What's the difference between the greenhouse effect and global warming?

Q3 Give two natural causes of climate change.

Q4 What's meant by 'carbon footprint'?

Exam Questions

Q1 a) Name three greenhouse gases. [2 marks]

b) Explain how greenhouse gases keep the temperature in the lower layer of the Earth's atmosphere higher than it would otherwise be. [3 marks]

c) What factors affect the contribution a gas makes to the greenhouse effect? [2 marks]

Q2 The concentration of carbon dioxide in the Earth's atmosphere has increased over the last 50 years.

a) Describe why using petrol as a fuel is not carbon neutral. [1 mark]

b) Bioethanol is a renewable fuel that is a viable alternative to petrol. It can be made from sugar cane.

(i) What process converts the sugar from the sugar cane into ethanol? [1 mark]

(ii) Explain why the use of bioethanol is considered to be carbon neutral. [2 marks]

Global Warming probably just isn't funny...

You may be sick of global warming, because it's all over the news these days. Well, tough — just think of all those poor, seasick chemists hauling bucketfuls of water out of the ocean and sticking litmus paper in to test its acidity (that's not actually what they do, clearly, but there is a lot of careful measuring involved to monitor what's changing and how).

Mass Spectra and Infrared Spectra

Get ready for the thrilling climax of the book — and watch out for the twist at the end...

Mass Spectrometry Can Help to Identify Compounds

1) You saw on pages 26-27 how **mass spectrometry** can be used to find **relative isotopic masses**, the **abundance** of different isotopes, and the **relative molecular mass**, M_r, of a compound.
2) Remember — to find the relative molecular mass of a compound you look at the **molecular ion peak** (the **M peak**) on the spectrum. Molecular ions are formed when molecules have **electrons** knocked off. The mass/charge value of the molecular ion peak is the molecular mass.

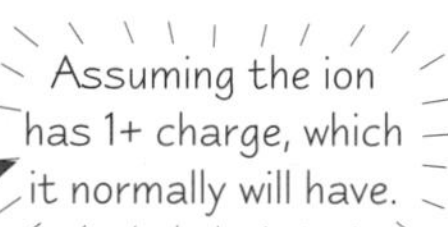

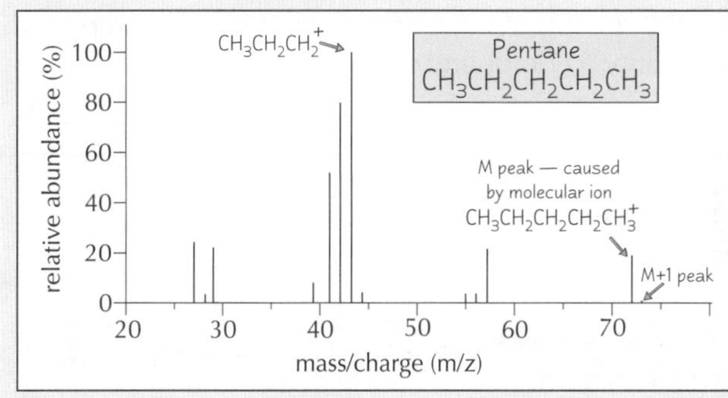

Here's the mass spectrum of pentane. Its M peak is at 72 — so the compound's M_r is 72.

For most <u>organic compounds</u> the M peak is the one with the second highest mass/charge ratio. The smaller peak to the right of the M peak is called the M+1 peak — it's caused by the presence of the carbon isotope ^{13}C (you don't need to worry about this at AS).

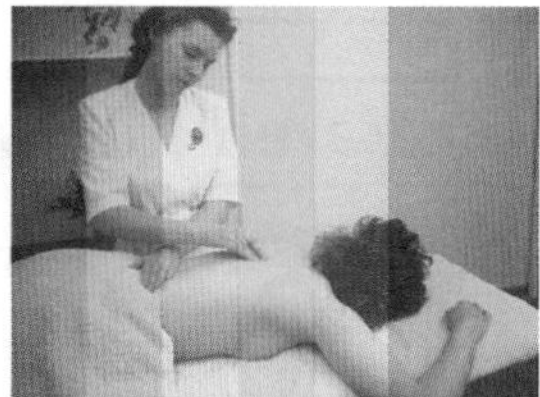

A massage spectrum

The **Molecular Ion** can be **Broken** into **Smaller Fragments**

The bombarding electrons make some of the molecular ions break up into **fragments**.
The fragments that are ions show up on the mass spectrum, making a **fragmentation pattern**. Fragmentation patterns are actually pretty cool because you can use them to identify **molecules** and even their **structure**.

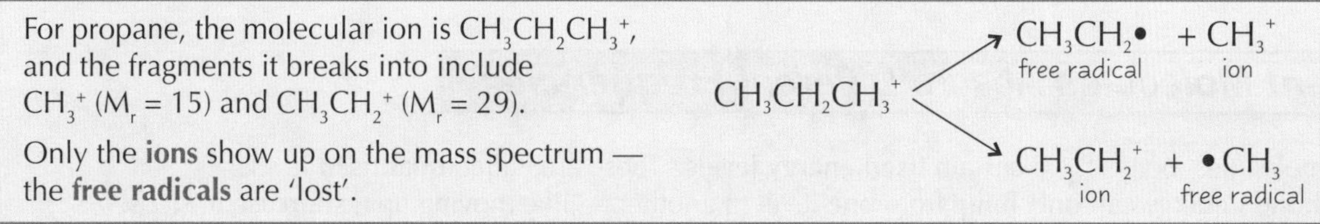

For propane, the molecular ion is $CH_3CH_2CH_3^+$, and the fragments it breaks into include CH_3^+ (M_r = 15) and $CH_3CH_2^+$ (M_r = 29).

Only the **ions** show up on the mass spectrum — the **free radicals** are 'lost'.

To work out the structural formula, you've got to work out what **ion** could have made each peak from its **m/z value**. (You assume that the m/z value of a peak matches the **mass** of the ion that made it.)

Example: Use this mass spectrum to work out the structure of the molecule:

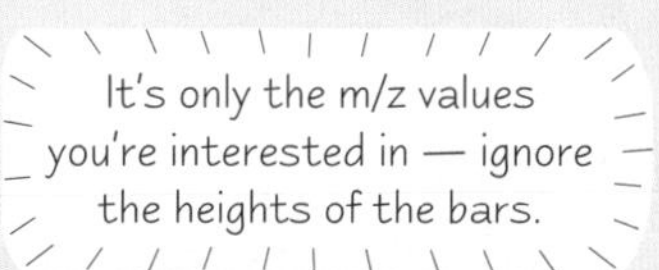

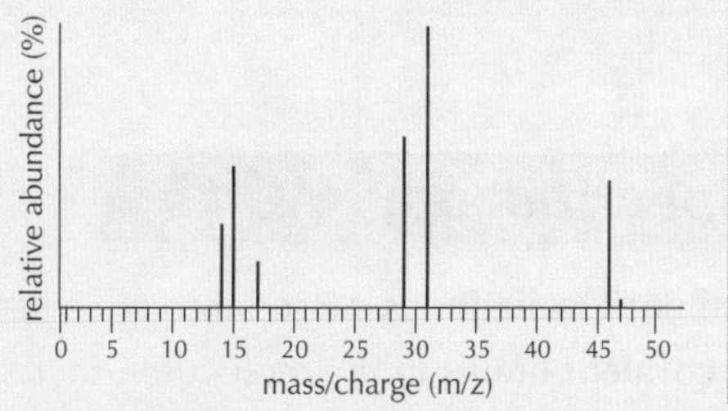

Fragment	Molecular Mass
CH_3	15
C_2H_5	29
C_3H_7	43
OH	17

1. Identify the fragments

This molecule's got a peak at 15 m/z, so it's likely to have a **CH_3 group**.

It's also got a peak at 17 m/z, so it's likely to have an **OH group**.

Other ions are matched to the peaks here:

31 CH_2OH^+; 29 $CH_2CH_3^+$; 15 CH_3^+; 14 CH_2^+; 17 OH^+; 46 M peak $CH_3CH_2OH^+$

relative abundance (%); mass/charge (m/z): 0, 5, 10, 15, 20, 25, 30, 35, 40, 45, 50

2. Piece them together to form a molecule with the correct M_r

Ethanol has all the fragments on this spectrum.

```
   H   H
   |   |
H—C—C—O—H
   |   |
   H   H
```

Ethanol's **molecular mass** is 46.
This should be the same as the m/z value of the M peak — it is.

Mass Spectra and Infrared Spectra

*Mass Spectrometry is Used to **Differentiate** Between **Similar Molecules***

1) Even if two **different compounds** contain **the same atoms**, you can still tell them apart with mass spectrometry because they won't produce exactly the same set of fragments.
2) The formulas of **propanal** and **propanone** are shown on the right. They've got the same M_r, but different structures, so they produce some **different fragments**. For example, propanal will have a C_2H_5 fragment but propanone won't.

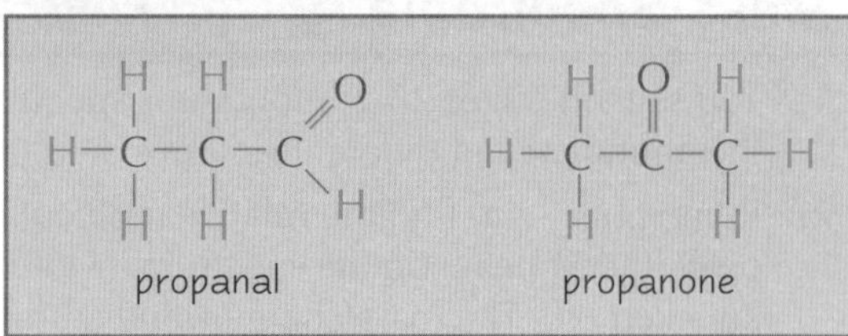

3) Every compound produces a different mass spectrum — so the spectrum's like a **fingerprint** for the compound. Large computer **databases** of mass spectra can be used to identify a compound from its spectrum.

Infrared Radiation** Makes Some **Bonds Vibrate More

1) Some molecules absorb energy from **infrared radiation**. The extra energy makes their covalent bonds **vibrate** more.
2) Only molecules made of **different atoms** can absorb infrared radiation. This is because the **polarities** of their bonds change as they vibrate.
3) So for example, oxygen (O_2) and nitrogen (N_2) don't absorb infrared radiation, but **carbon dioxide**, **water**, **nitric oxide (NO)** and **methane** do.
4) Gases that **do** absorb infrared radiation are called **greenhouse gases** because they stop some of the radiation emitted by the Earth from escaping into space (see page 112).

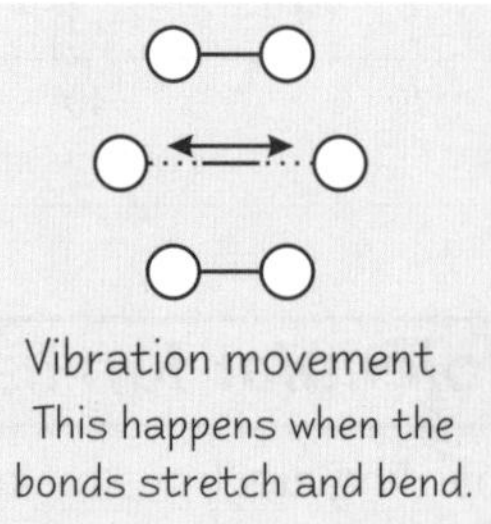
Vibration movement
This happens when the bonds stretch and bend.

Different Molecules** Absorb Different **Frequencies

- Gas molecules' bonds have **certain fixed energy levels**. These are called **quantised** levels. So a bond's energy can only **jump** from one level to another — like moving up a **staircase** in steps.
- This means that only frequencies of radiation corresponding to particular amounts of energy are absorbed. **Different molecules** absorb **different frequencies** of radiation.

*...Which is How **Infrared Spectroscopy** Works*

1) In infrared (IR) spectroscopy, a beam of **IR radiation** is passed through a sample of a chemical.
2) The IR radiation is absorbed by the **covalent bonds** in the molecules, increasing their **vibrational** energy, as above.
3) **Bonds between different atoms** absorb **different frequencies** of IR radiation. Bonds in different **places** in a molecule absorb different frequencies too — so the O–H group in an **alcohol** and the O–H in a **carboxylic acid** absorb different frequencies. This table shows what **frequencies** different bonds absorb:

Functional group	Where it's found	Frequency/ Wavenumber (cm^{-1})	Type of absorption
C–H	most organic molecules	2800 - 3100	strong, sharp
O–H	alcohols	3200 - 3550	strong, broad
O–H	carboxylic acids	2500 - 3300	medium, broad
N–H	amines (e.g. methylamine, CH_3NH_2)	3200 - 3500	strong, sharp
C=O	aldehydes, ketones, carboxylic acids	1680 - 1750	strong, sharp
C–X	haloalkanes	500 - 1000	strong, sharp

This tells you what the peak on the graph will look like.

You don't need to learn this data, but you do need to understand how to use it.

Mass Spectra and Infrared Spectra

Infrared Spectroscopy *Helps You* ***Identify Organic Molecules***

An infrared spectrometer produces a **graph** that shows you what frequencies of radiation the molecules are absorbing. So you can use it to identify the **functional groups** in a molecule:

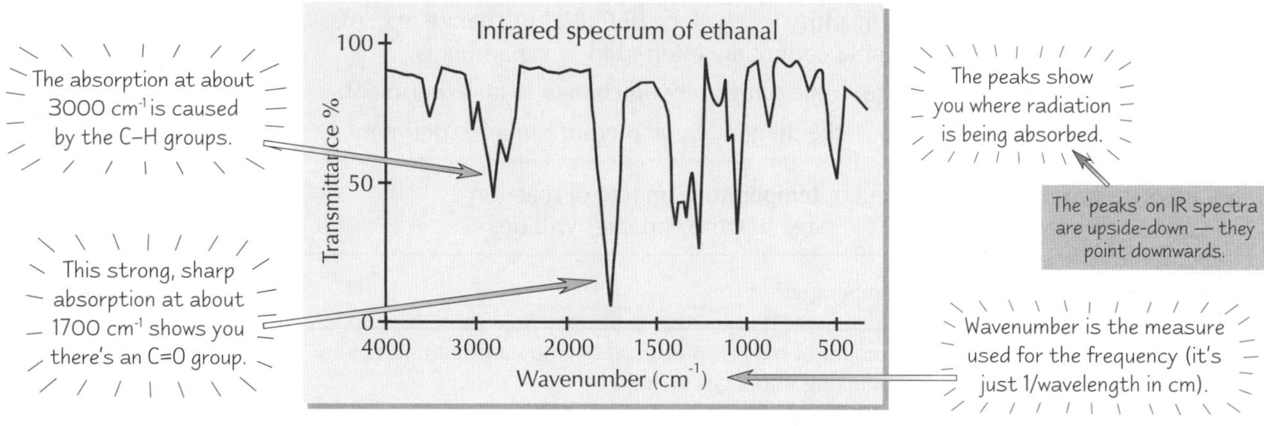

This also means that you can tell if a functional group has **changed** during a reaction. For example, if you **oxidise** an **alcohol** to an **aldehyde** you'll see the O–H absorption **disappear** from the spectrum, and a C=O absorption **appear**.

Practice Questions

Q1 In mass spectrometry, what is meant by: a) the molecular ion? b) the M peak?

Q2 How do fragments get formed?

Q3 Why don't all molecules absorb infrared energy?

Q4 Why do most infrared spectra of organic molecules have a strong, sharp peak at around 3000 cm^{-1}?

Exam Questions

Q1 To the right is the mass spectrum of an organic compound, Q.

a) What is the M_r of compound Q? [1 mark]

b) What fragments are the peaks marked X and Y most likely to correspond to? [2 marks]

c) Suggest a structure for this compound. [1 mark]

d) Why is it unlikely that this compound is an alcohol? [2 marks]

relative abundance
Y
X
0 5 10 15 20 25 30 35 40 45
mass/charge (m/z)

Q2 A molecule with a molecular mass of 74 produces the following IR spectrum.

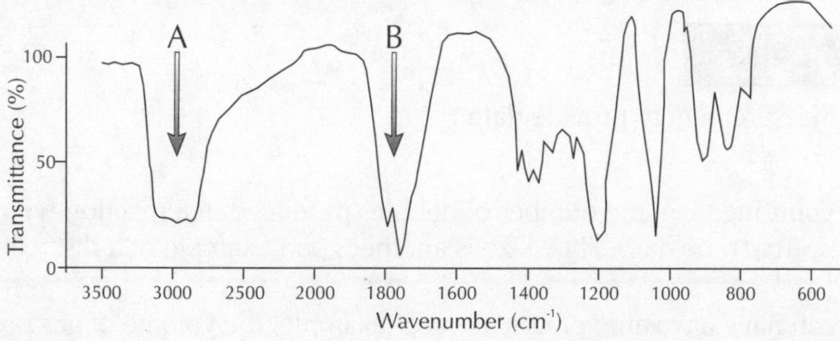

Use the infrared absorption data on the opposite page.

a) Which functional groups are responsible for peaks A and B? [2 marks]

b) Suggest the molecular formula and name of this molecule. Explain your answer. [3 marks]

I wonder what the infrared spectrum of a fairy cake would look like...

I don't suppose I'll ever know. Very squiggly I imagine. You don't have to remember what any of the graphs look like. But you need to know how to interpret them, because sure as eggs are eggs it'll be in the exam. Don't worry, I haven't forgotten I said there was twist at the end... erm... hydrogen was my sister all along... and all the elements went to live in Jamaica. The End.

Practical and Investigative Skills

You're going to have to do some practical work too — and once you've done it, you have to make sense of your results...

*Make it a **Fair Test** — Control your **Variables***

You probably know this all off by heart but it's easy to get mixed up sometimes. So here's a quick recap:

Variable — A variable is a **quantity** that has the **potential to change**, e.g. mass.
There are two types of variable commonly referred to in experiments:

- **Independent variable** — the thing that you **change** in an experiment.
- **Dependent variable** — the thing that you **measure** in an experiment.

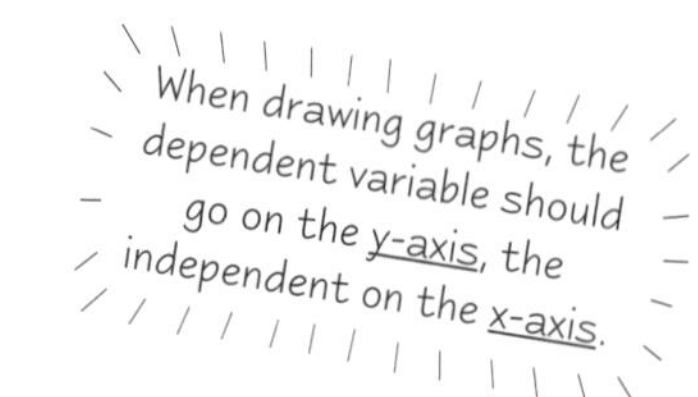

So, if you're investigating the effect of **temperature** on rate of reaction using the apparatus on the right (see page 92), the variables will be:

Independent variable	Temperature
Dependent variable	Amount of oxygen produced — you can measure this by collecting it in a gas syringe
Other variables — you MUST keep these the same	Concentration and volume of solutions, mass of solids, pressure, the presence of a catalyst and the surface area of any solid reactants

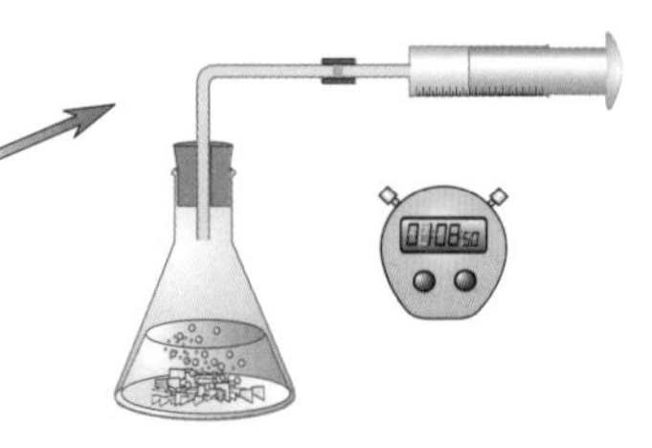

*Organise Your Results in a **Table** — And Watch Out For **Anomalous** Ones*

Before you start your experiment, make a **table** to write your results in.
You'll need to repeat each test at least three times to check your results are reliable.

This is the sort of table you might end up with when you investigate the effect of **temperature** on **reaction rate**. (You'd then have to do the same for **different temperatures**.)

Temperature	Time (s)	Volume of gas evolved (cm^3) Run 1	Volume of gas evolved (cm^3) Run 2	Volume of gas evolved (cm^3) Run 3	Average volume of gas evolved (cm^3)
20 °C	**10**	8	7	8	**7.7**
	20	17	19	20	**18.7**
	30	28	20	30	**29**

Find the average of each set of repeated values.
You need to add them all up and divide by how many there are.
E.g.: $(8 + 7 + 8) \div 3 = 7.7$ cm^3

Watch out for **anomalous results**. These are ones that don't fit in with the other values and are likely to be wrong. They're likely to be due to random errors — here the syringe plunger may have got stuck.
Ignore anomalous results when you calculate the average.

*Know Your Different Sorts of **Data***

Experiments always involve some sort of measurement to provide **data**.
There are different types of data —

Discrete — you get discrete data by **counting**. E.g. the number of bubbles produced in a reaction would be discrete. You can't have 1.25 bubbles. That'd be daft. Shoe size is another good example of a discrete variable.

Continuous — a continuous variable can have **any value** on a scale. For example, the volume of gas produced or the mass of products from a reaction. You can never measure the exact value of a continuous variable.

Categoric — a categoric variable has values that can be sorted into **categories**. For example, the colours of solutions might be blue, red and green. Or types of material might be wood, steel, glass.

Ordered (ordinal) — Ordered data is similar to categoric, but the categories can be **put in order**. For example, if you classify reactions as 'slow', 'fairly fast' and 'very fast' you'd have ordered data.

Practical and Investigative Skills

Graphs: *Line, Bar or Scatter — Use the Best Type*

You'll usually be expected to make a **graph** of your results. Not only are graphs **pretty**, they make your data **easier to understand** — so long as you choose the right type.

Line graphs are best when you have **two sets of continuous data**. For example:

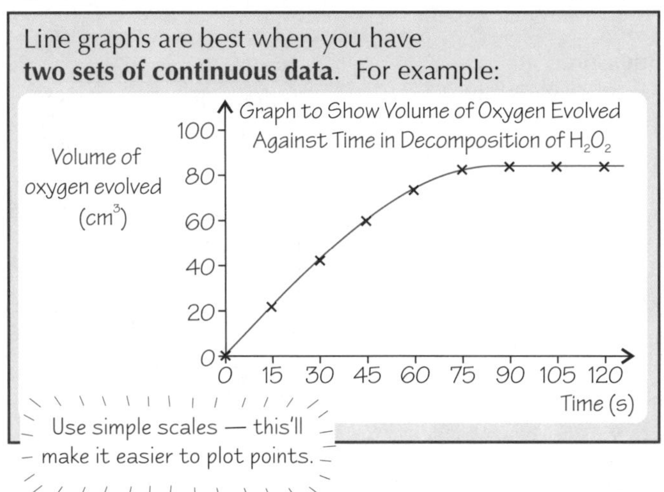

Use simple scales — this'll make it easier to plot points.

You should use a bar chart when one of your data sets is **categoric or ordered data**. For example:

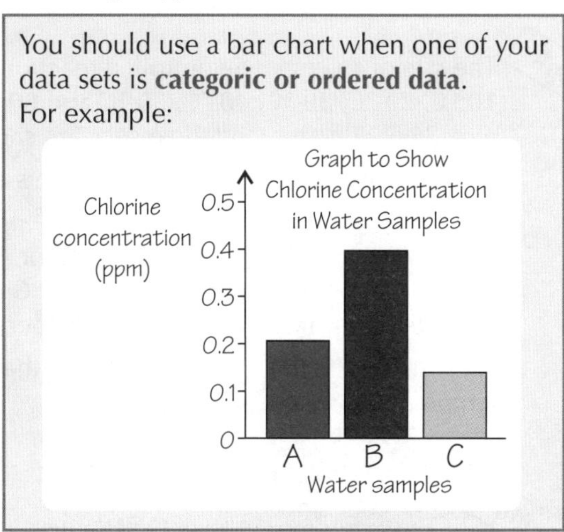

Scatter plots are great for showing how two sets of **discrete or continuous data** are related (or **correlated**).

Don't try to join all the points — draw a **line of best fit** to show the **trend**.

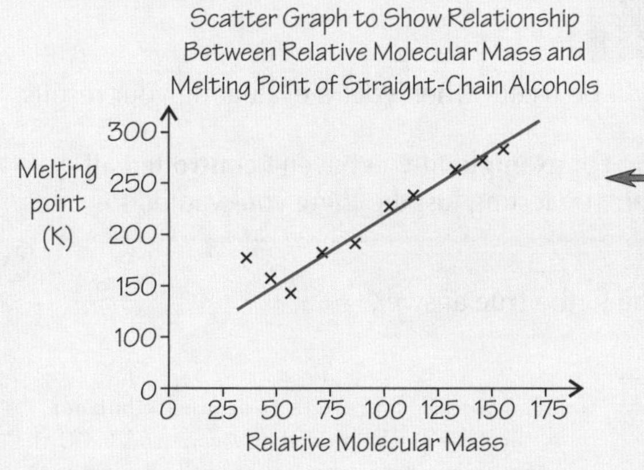

Scatter Graphs Show The Relationship Between Variables

Correlation describes the **relationship** between two variables — the independent one and the dependent one.

Data can show:

1) **Positive correlation** — as one variable **increases** the other **increases**. The graph on the left shows positive correlation.
2) **Negative correlation** — as one variable **increases** the other **decreases**.
3) **No correlation** — there is **no relationship** between the two variables.

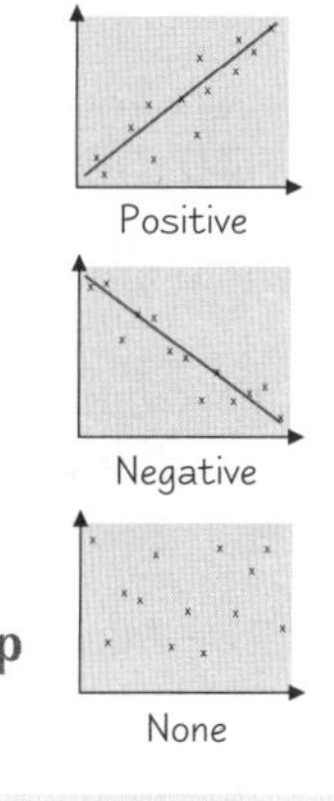

There's also pie charts. These are normally used to display categoric data.

Whatever type of graph you make, you'll ONLY get full marks if you:

- Choose a sensible scale — don't do a tiny graph in the corner of the paper.
- Label both axes — including units.
- Plot your points accurately — using a sharp pencil.

*Correlation **Doesn't** Mean **Cause** — Don't Jump to Conclusions*

1) Ideally, only **two** quantities would **ever** change in any experiment — everything else would remain **constant**.
2) But in experiments or studies outside the lab, you **can't** usually control all the variables. So even if two variables are correlated, the change in one may **not** be causing the change in the other. Both changes might be caused be a **third variable**.

Watch out for bias too — for instance, a bottled water company might point these studies out to people without mentioning any of the doubts.

Example

For example: Some studies have found a correlation between **drinking chlorinated tap water** and the risk of developing certain cancers. So some people argue that this means water shouldn't have chlorine added.

BUT it's hard to control all the variables between people who drink tap water and people who don't. It could be many lifestyle factors.

Or, the cancer risk could be affected by something else in tap water — or by whatever the non-tap water drinkers drink instead...

Practical and Investigative Skills

Don't Get **Carried Away** When Drawing Conclusions

The **data** should always **support** the conclusion. This may sound obvious but it's easy to **jump** to conclusions. Conclusions have to be **specific** — not make sweeping generalisations.

Example

The rate of an enzyme-controlled reaction was measured at **10 °C, 20 °C, 30 °C, 40 °C, 50 °C and 60 °C**. All other variables were kept constant, and the results are shown in this graph.

A science magazine **concluded** from this data that enzyme X works best at **40 °C**. The data **doesn't** support this.

The enzyme **could** work best at 42 °C or 47 °C but you can't tell from the data because **increases** of **10 °C** at a time were used. The rate of reaction at in-between temperatures **wasn't** measured.

All you know is that it's faster at **40 °C** than at any of the other temperatures tested.

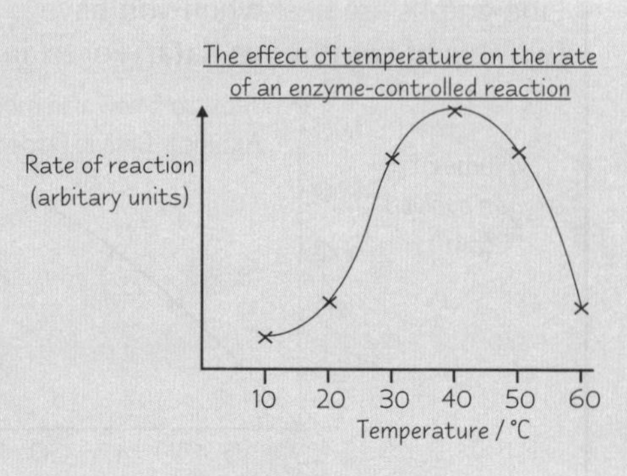

Example

The experiment above **ONLY** gives information about this particular enzyme-controlled reaction. You can't conclude that **all** enzyme-controlled reactions happen faster at a particular temperature — only this one. And you can't say for sure that doing the experiment at, say, a different constant pressure, wouldn't give a different optimum temperature.

You need to Look **Critically** at Your Results

There are a few bits of lingo that you need to understand. They'll be useful when you're evaluating your results.

1) **Valid results** — Valid results answer the original question. For example, if you haven't **controlled all the variables** your results won't be valid, because you won't be testing just the thing you wanted to.

2) **Accurate** — Accurate results are those that are **really close** to the **true** answer.

3) **Precise results** — These are results taken using **sensitive instruments** that measure in **small increments**, e.g. pH measured with a meter (pH 7.692) will be **more precise** than pH measured with paper (pH 7).

It's possible for results to be precise **but not** accurate, e.g. a balance that weighs to 1/1000 th of a gram will give precise results but if it's not **calibrated** properly the results won't be accurate.

You may have to calculate the **percentage uncertainty** of a measurement — that's how much error it might have. E.g. the uncertainty associated with a pipette is 0.06 cm³. If you measure 25 cm³ with it, the percentage uncertainty is:

$$\text{percentage uncertainty} = \frac{\text{uncertainty}}{\text{reading}} \times 100 = \frac{0.06}{25} \times 100 = \mathbf{0.24\%}$$

See pages 86-87 for more on uncertainty.

Watch out when you're using burettes in titrations though. You take two readings, an initial one and a final one — the titre is the final volume minus the initial volume. If each reading has an uncertainty of 0.05 cm³ associated with it, the **overall uncertainty** for the titre is 0.05 × 2 = 0.1 cm³.

4) **Reliable results** — **Reliable** means the results can be **consistently reproduced** in independent experiments. And if the results are reproducible they're more likely to be **true**. If the data isn't reliable for whatever reason you **can't draw** a valid **conclusion**.

For experiments, the **more repeats** you do, the **more reliable** the data. If you get the **same result** twice, it could be the correct answer. But if you get the same result **20 times**, it'd be much more reliable. And it'd be even more reliable if everyone in the class got about the same results using different apparatus.

Practical and Investigative Skills

Organic Chemistry Uses some Specific Techniques

There are some **practical techniques** that get used a lot in organic chemistry. The products from organic reactions are often **impure** — so you've got to know how to get rid of the unwanted by-products or leftover reactants.

The method for turning an **alcohol** into a **chloroalkane** (see page 98) is a really useful example because it involves quite a few of those techniques.

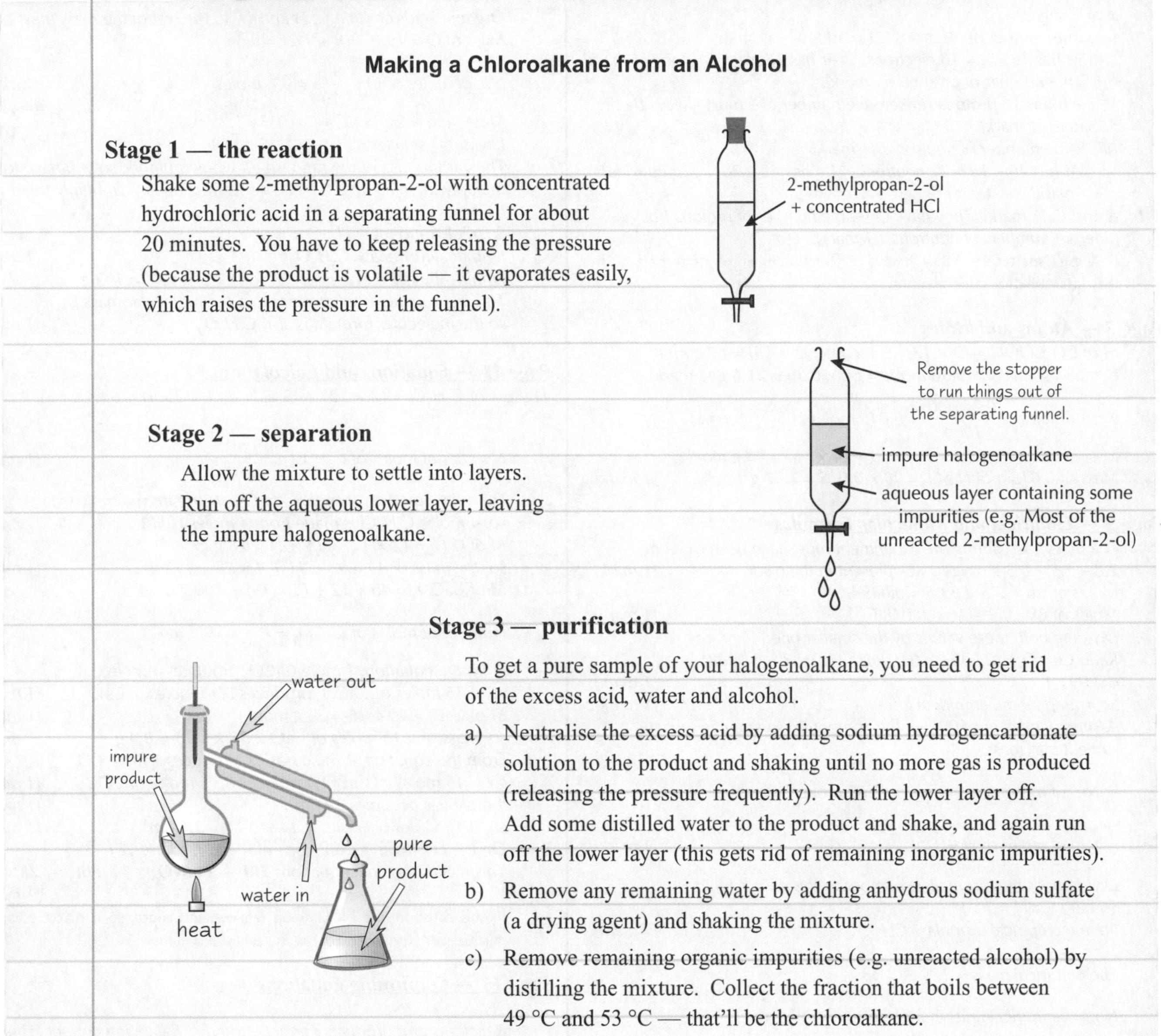

Making a Chloroalkane from an Alcohol

Stage 1 — the reaction

Shake some 2-methylpropan-2-ol with concentrated hydrochloric acid in a separating funnel for about 20 minutes. You have to keep releasing the pressure (because the product is volatile — it evaporates easily, which raises the pressure in the funnel).

Stage 2 — separation

Allow the mixture to settle into layers. Run off the aqueous lower layer, leaving the impure halogenoalkane.

Stage 3 — purification

To get a pure sample of your halogenoalkane, you need to get rid of the excess acid, water and alcohol.

a) Neutralise the excess acid by adding sodium hydrogencarbonate solution to the product and shaking until no more gas is produced (releasing the pressure frequently). Run the lower layer off. Add some distilled water to the product and shake, and again run off the lower layer (this gets rid of remaining inorganic impurities).

b) Remove any remaining water by adding anhydrous sodium sulfate (a drying agent) and shaking the mixture.

c) Remove remaining organic impurities (e.g. unreacted alcohol) by distilling the mixture. Collect the fraction that boils between 49 °C and 53 °C — that'll be the chloroalkane.

Work Safely and Ethically — Don't Blow Up the Lab or Harm Small Animals

In any experiment you'll be expected to show that you've thought about the **risks and hazards** (see page 57 for more.) It's generally a good thing to wear a lab coat and safety glasses, but you may need to take additional safety measures, depending on the experiment. For example, anything involving nasty gases will need to be done in a fume cupboard. Also, check the **hazard symbols** on any chemicals you're using (see page 57).

You need to make sure you're working **ethically** too. This is most important if there are other people or animals involved. You have to put their welfare first.

Answers

Unit 1: Section 1 — Formulas and Equations

Page 5 — The Atom

1) a) Similarity — They've all got the same number of protons/electrons. [1 mark]
Difference — They all have different numbers of neutrons. [1 mark]
b) 1 proton [1 mark], 1 neutron (2 – 1) [1 mark], 1 electron [1 mark].
c) ^3H. [1 mark]
Since tritium has 2 neutrons in the nucleus and also 1 proton, it has a mass number of 3. You could also write ^3_1H but you don't really need the atomic number.

2) a) (i) Same number of electrons. [1 mark]
$^{32}_{16}S^{2-}$ has 16 + 2 = 18 electrons. $^{40}_{18}Ar$ has 18 electrons too. [1 mark]
(ii) Same number of protons. [1 mark].
Each has 16 protons (the atomic number of S must always be the same) [1 mark].
(iii) Same number of neutrons. [1 mark]
$^{40}_{18}Ar$ has 40 – 18 = 22 neutrons. $^{42}_{20}Ca$ has 42 – 20 = 22 neutrons. [1 mark]
b) **A** and **C**. [1 mark] They have the same number of protons but different numbers of neutrons. [1 mark].
It doesn't matter that they have a different number of electrons because they are still the same element.

Page 7 — Atoms and Moles

1) M of CH_3COOH = (2 × 12) + (4 × 1) + (2 × 16) = 60 g mol^{-1}
[1 mark] so mass of 0.36 moles = 60 × 0.36 = **21.6 g** [1 mark]

2) No. of moles = $\frac{0.25 \times 60}{1000}$ = 0.015 moles H_2SO_4 [1 mark]
M of H_2SO_4 = (2 × 1) + (1 × 32) + (4 × 16) = 98 g mol^{-1}
Mass of 0.015 mol H_2SO_4 = 98 × 0.015 = **1.47 g** [1 mark]

Page 9 — Empirical and Molecular Formulas

1) The mass 'lost' during the experiment must have been oxygen.
2.8 – 2.5 = 0.3 g oxygen was present in the oxide. [1 mark]
Moles of Cu = 2.5 ÷ 63.5 = 0.0394.
Moles of O = 0.3 ÷ 16 = 0.0188 [1 mark]
Dividing both these values by the smaller one:
Ratio Cu : O = (0.0394 ÷ 0.0188) : (0.0188 ÷ 0.0188)
= 2.09 : 1 [1 mark]
So, rounding off, empirical formula = Cu_2O [1 mark]

2) Assume you've got 100 g of the compound so you can turn the % straight into mass.

No. of moles of C = $\frac{92.3}{12}$ = 7.69 moles

No. of moles of H = $\frac{7.7}{1}$ = 7.7 moles [1 mark]

Divide both by the smallest number, in this case 7.69.
So ratio C : H = 1 : 1
So, the empirical formula = CH [1 mark]

The empirical mass = 12 + 1 = 13

No. of empirical units in molecule = $\frac{78}{13}$ = 6

So the molecular formula = $\mathbf{C_6H_6}$ [1 mark]

3) The magnesium is burning, so it's reacting with oxygen and the product is magnesium oxide.
First work out the number of moles of each element.

No. of moles Mg = $\frac{1.2}{24}$ = 0.05 moles

Mass of O is everything that isn't Mg: 2 – 1.2 = 0.8 g

No. of moles O = $\frac{0.8}{16}$ = 0.05 moles [1 mark]

Ratio Mg : O = 0.05 : 0.05
Divide both by the smallest number, in this case 0.05.
So ratio Mg : O = 1 : 1
So the empirical formula is **MgO** [1 mark]

4) First calculate the no. of moles of each product and then the mass of C and H:

No. of moles of CO_2 = $\frac{33}{44}$ = 0.75 moles

Mass of C = 0.75 × 12 = 9 g

No. of moles of H_2O = $\frac{10.8}{18}$ = 0.6 moles

0.6 moles H_2O = 1.2 moles H
Mass of H = 1.2 × 1 = 1.2 g [1 mark]
Organic acids contain C, H and O, so the rest of the mass must be O.
Mass of O = 19.8 – (9 + 1.2) = 9.6 g

No. of moles of O = $\frac{9.6}{16}$ = 0.6 moles [1 mark]

Mole ratio = C : H : O = 0.75 : 1.2 : 0.6
Divide by smallest 1.25 : 2 : 1
The carbon part of the ratio isn't a whole number, so you have to multiply them all up until it is. As its fraction is ¼, multiply them all by 4.
So, mole ratio = C : H : O = 5 : 8 : 4
Empirical formula = $\mathbf{C_5H_8O_4}$ [1 mark]
Empirical mass = (12 × 5) + (1 × 8) + (16 × 4) = 132 g
This is the same as what we're told the molecular mass is, so the molecular formula is also $\mathbf{C_5H_8O_4}$. [1 mark]

Page 11 — Equations and Calculations

1) M of C_2H_5Cl = (2 × 12) + (5 × 1) + (1 × 35.5) = 64.5 g mol^{-1} [1 mark]

Number of moles of C_2H_5Cl = $\frac{258}{64.5}$ = 4 moles [1 mark]

From the equation, 1 mole C_2H_5Cl is made from 1 mole C_2H_4
so, 4 moles C_2H_5Cl is made from 4 moles C_2H_4. [1 mark]
M of C_2H_4 = (2 × 12) + (4 × 1) = 28 g mol^{-1}
so, the mass of 4 moles C_2H_4 = 4 × 28 = **112 g** [1 mark]

2) a) M of $CaCO_3$ = 40 + 12 + (3 × 16) = 100 g mol^{-1}

Number of moles of $CaCO_3$ = $\frac{15}{100}$ = 0.15 moles

From the equation, 1 mole $CaCO_3$ produces 1 mole CaO
so, 0.15 moles of $CaCO_3$ produces 0.15 moles of CaO. [1 mark]
M of CaO = 40 + 16 = 56 g mol^{-1} [1 mark]
so, mass of 0.15 moles of CaO = 56 × 0.15 = **8.4 g** [1 mark]
b) From the equation, 1 mole $CaCO_3$ produces 1 mole CO_2
so, 0.15 moles of $CaCO_3$ produces 0.15 moles of CO_2. [1 mark]
1 mole gas occupies 24 dm^3, [1 mark]
so, 0.15 moles occupies = 24 × 0.15 = **3.6 dm^3** [1 mark]

3) On the LHS, you need 2 each of K and I, so use 2KI
This makes the final equation: $\mathbf{2KI + Pb(NO_3)_2 \rightarrow PbI_2 + 2KNO_3}$ [1 mark]
In this equation, the NO_3 group remains unchanged, so it makes balancing much easier if you treat it as one indivisible lump.

Page 13 — Confirming Equations

1) a) $2NaN_3 \rightarrow 2Na + 3N_2$
Reactants and products correct [1 mark]. Balancing correct [1 mark].
b) (i) M of NaN_3 = 23 + (3 × 14) = 65 g mol^{-1} [1 mark]
0.325/65 = **0.005 moles** [1 mark]
(ii) 180/24000 = **0.0075 moles** [1 mark]
(iii) ratio NaN_3:N_2 = 0.005:0.0075 [1 mark] = **2:3** [1 mark]

2) a) By delivering the gas to an upturned measuring cylinder (or burette) [1 mark] filled with water and displacing the water [1 mark]
(Alternatively: connecting the conical flask to a gas syringe to measure the volume produced [2 marks])
b) E.g. Not all of the hydrogen gas may be collected (some may escape). / The equipment used to measure the mass of magnesium and that used to measure the volume of hydrogen will have a limited precision. / Gas may not be at r.t.p. when volume measured.
[1 mark for each sensible reason, up to a maximum of 2 marks.]

Answers

Page 15 — Making Salts

1) a) no. moles = 0.2 mol dm^{-3} × 0.05 dm^3 = **0.01 moles** [1 mark]
(50 cm^3 = 0.05 dm^3)

b) molar mass of the hydrated copper(II) sulfate
= 63.5 + 32 + (4 × 16) + (5 × (2 + 16)) = 249.5 g mol^{-1} [1 mark]
ratio of sulfuric acid to copper sulfate = 1:1 (from balanced equation)
so max. no. moles copper sulfate = 0.01 [1 mark]
mass of 0.01 moles of hydrated copper sulfate
= 0.01 × 249.5 = 2.495 g [1 mark]
percentage yield = (1.964/2.495) × 100 = **78.72%** [1 mark]

Page 17 — Atom Economy and Percentage Yield

1) a) 2 is an addition reaction [1 mark]

b) For reaction 1: % atom economy
= $M_r(C_2H_5Cl) \div [M_r(C_2H_5Cl) + M_r(POCl_3) + M_r(HCl)]$ [1 mark]
= [(2 × 12) + (5 × 1) + 35.5]
÷ [(2 × 12) + (5 × 1) + 35.5 + 31 + 16 + (3 × 35.5) + 1 + 35.5] × 100 [1 mark]
= (64.5 ÷ 254.5) × 100 = 25.3% [1 mark]

c) The atom economy is 100% because there is only one product (there are no by-products) [1 mark].

2) a) There is only one product, so the theoretical yield can be calculated by adding the masses of the reactants [1 mark].
So theoretical yield = 0.275 + 0.142 = 0.417 g [1 mark]

b) percentage yield = (0.198 ÷ 0.417) × 100 = 47.5% [1 mark]

c) Changing reaction conditions will have no effect on atom economy [1 mark]. Since the equation shows that there is only one product, the atom economy will always be 100% [1 mark].
Atom economy is related to the type of reaction — addition, substitution, etc. — not to the quantities of products and reactants.

Unit 1: Section 2 — Energetics

Page 19 — Enthalpy Changes

1)

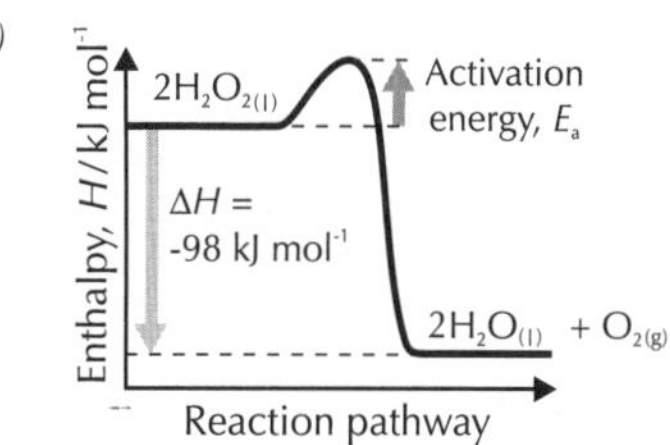

Reactants lower in energy than products [1 mark]. Activation energy correctly labelled [1 mark]. ΔH correctly labelled with arrow pointing **downwards** [1 mark].
For an exothermic reaction, the ΔH arrow points downwards, but for an endothermic reaction it points upwards. The activation energy arrow always points upwards though.

2) a) $CH_3OH_{(l)} + 1\frac{1}{2}O_{2(g)} \rightarrow CO_{2(g)} + 2H_2O_{(l)}$
Correct balanced equation [1 mark]. Correct state symbols [1 mark].
It is perfectly OK to use halves to balance equations. Make sure that only 1 mole of CH_3OH is combusted, as it says in the definition for ΔH_c°.

b) $C_{(s)} + 2H_{2(g)} + \frac{1}{2}O_{2(g)} \rightarrow CH_3OH_{(l)}$
Correct balanced equation [1 mark]. Correct state symbols [1 mark].

c) Only 1 mole of C_3H_8 should be shown according to the definition of ΔH_c° [1 mark].
You really need to know the definitions of the standard enthalpy changes off by heart. There are loads of nit-picky little details they could ask you questions about.

Page 21 — Finding Enthalpy Changes

1) ΔT = 25.5 – 19 = 6.5 °C [1 mark]
m = 25 + 25 = 50 cm^3 of solution which has a mass of 50 g (assume density to be 1.0 g cm^{-3}) [1 mark]
Heat produced by reaction= mcΔT
= 50 × 4.18 × 6.5 = 1358.5 J [1 mark]

No. of moles of HCl = $\frac{1 \times 25}{1000}$ = 0.025 moles [1 mark]

0.025 moles of HCl produces 1358.5 J of heat, therefore 1 mole of HCl produces $\frac{1358.5}{0.025}$ [1 mark] = 54 340 J ≈ 54.3 kJ

So the enthalpy change is **–54.3 kJ mol^{-1}** [1 mark for correct number, 1 mark for minus sign].
You need the minus sign because it's exothermic.

2) No. of moles of $CuSO_4$ = $\frac{0.2 \times 50}{1000}$ = 0.01 moles [1 mark]

From the equation, 1 mole of $CuSO_4$ reacts with 1 mole of Zn.
So, 0.01 moles of $CuSO_4$ reacts with 0.01 moles of Zn [1 mark].
Heat produced by reaction= mcΔT
= 50 × 4.18 × 2.6 = 543.4 J [1 mark]
0.01 moles of zinc produces 543.4 J of heat, therefore 1 mole of zinc produces $\frac{543.4}{0.01}$ [1 mark] = 54 340 J ≈ 54.3 kJ

So the enthalpy change is **–54.3 kJ mol^{-1}** (you need the minus sign because it's exothermic) [1 mark for correct number, 1 mark for minus sign].
It'd be dead easy to work out the heat produced by the reaction, breathe a sigh of relief and sail on to the next question. But you need to find out the enthalpy change when 1 mole of zinc reacts. It's always a good idea to reread the question and check you've actually answered it.

Page 23 — Using Hess's Law

1) ΔH_r° = sum of ΔH_f°(products) – sum of ΔH_f°(reactants)
= [0 + (3 × –602)] [1 mark] – [–1676 + (3 × 0)] [1 mark]
= **–130 kJ mol^{-1}** [1 mark]
Don't forget the units. It's a daft way to lose marks.

2) ΔH_f° = ΔH_c°(glucose) – 2 × ΔH_c°(ethanol) [1 mark]
= [–2820] – [(2 × –1367)] [1 mark]
= **–86 kJ mol^{-1}** [1 mark]

Page 25 — Bond Enthalpy

1) a)

$N_{2(g)} + 3H_{2(g)} \xrightarrow{\Delta H_1} 2NH_{3(g)}$

ΔH_3 (from $N_{2(g)} + 3H_{2(g)}$), ΔH_2 (from $2NH_{3(g)}$) → $2N_{(g)} + 6H_{(g)}$

$\Delta H_1 = 2\Delta H_f^\circ(NH_{3(g)})$
$\Delta H_2 = 6E(N–H)$
$\Delta H_3 = 2\Delta H_{at}^\circ(N) + 6\Delta H_{at}^\circ(H)$
[1 mark for correctly drawing Hess cycle. 1 mark for correctly defining each of ΔH_1, ΔH_2 and ΔH_3. Award marks if all of the quantities are halved.]

b) $2\Delta H_f^\circ(NH_3)$ = (2 × +473) + (6 × +218) – (6 × +391) [1 mark]
$2\Delta H_f^\circ(NH_3)$ = +946 + 1308 – 2346 = –92 kJ mol^{-1} [1 mark]

$\Delta H_f^\circ(NH_3) = \frac{-92}{2}$ [1 mark] = –46 kJ mol^{-1} [1 mark]

Remember — there's 6 N–H bonds to be broken in $2NH_3$.

c) 391 kJ mol^{-1} is the average N–H bond energy in ammonia.
The data book value of N–H is an average of N–H bond energies in many molecules [1 mark], like amines and acid amides.

2) The C–H bond will probably break first [1 mark]. It has a lower mean bond enthalpy than the O=O bond [1 mark].

Answers

Unit 1: Section 3 — Atomic Structure

Page 27 — Mass Spectrometry

1) a) *First multiply each relative abundance by the relative mass —*
120.8 × 63 = 7610.4, 54.0 × 65 = 3510.0
Next add up the products —
7610.4 + 3510.0 = 11 120.4 *[1 mark]*
Now divide by the total abundance (120.8 + 54.0 = 174.8)

$$A_r(Cu) = \frac{11120.4}{174.8} \approx \mathbf{63.6} \qquad [1\ mark]$$

You can check your answer by seeing if A_r(Cu) is in between 63 and 65 (the lowest and highest relative isotopic masses).

b) *A sample of copper is a mixture of 2 isotopes of different abundances [1 mark]. The weighted average mass of these isotopes isn't a whole number [1 mark].*

2) a) *Mass spectrometry. [1 mark]*

b) *You use pretty much the same method here as for question 1)a).*
93.11 × 39 = 3631.29, 0.12 × 40 = 4.8, 6.77 × 41 = 277.57
3631.29 + 4.8 + 277.57 = 3913.66 *[1 mark]*
This time you divide by 100 because they're percentages.

$$A_r(K) = \frac{3913.66}{100} \approx \mathbf{39.14} \qquad [1\ mark]$$

Again check your answer's between the lowest and highest relative isotopic masses, 39 and 41. A_r(K) is closer to 39 because most of the sample (93.11%) is made up of this isotope.

Page 29 — Electronic Structure

1) a) *K atom: $1s^2\ 2s^2\ 2p^6\ 3s^2\ 3p^6\ 4s^1$ [1 mark]*
K^+ ion: $1s^2\ 2s^2\ 2p^6\ 3s^2\ 3p^6$ [1 mark]

b) Oxygen electron Configuration: 1s [↑↓] 2s [↑↓] 2p [↑↓][↑][↑]

[1 mark for the correct number of electrons in each sub-shell. 1 mark for having spin-pairing in one of the p orbitals and parallel spins in the other two p orbitals. A box filled with 2 arrows is spin pairing — 1 up and 1 down. If you've put the four p electrons into just 2 orbitals, it's wrong.]

c) *The outer shell electrons in potassium and oxygen can get close to the outer shells of other atoms, so they can be transferred or shared [1 mark]. The inner shell electrons are tightly held and shielded from the electrons in other atoms/molecules [1 mark].*

2) a) *$1s^2\ 2s^2\ 2p^6\ 3s^2\ 3p^6\ 3d^5\ 4s^2$. [1 mark]*

b) *Germanium ($1s^2\ 2s^2\ 2p^6\ 3s^2\ 3p^6\ 3d^{10}\ 4s^2\ 4p^2$). [1 mark].*
(The 4p sub-shell is partly filled, so it must be a p block element.)

c) *Ar (atom) [1 mark], K^+ (positive ion) [1 mark], Cl^- (negative ion) [1 mark]. You also could have suggested Ca^{2+}, S^{2-} or P^{3-}.*

d) Al^{3+} electron Configuration: 1s [↑↓] 2s [↑↓] 2p [↑↓][↑↓][↑↓]

[1 mark for the correct number of electrons in each sub-shell. 1 mark for one arrow in each box pointing up, and one pointing down.]

Page 31 — Ionisation Energies

1) a) $C_{(g)} \rightarrow C^+_{(g)} + e^-$
[1 mark for the correct equation. 1 mark if both state symbols show gaseous state.]

b) *First ionisation energy generally increases as nuclear charge increases [1 mark].*

c) *As the nuclear charge increases there is a stronger force of attraction between the nucleus and the electron [1 mark] and so more energy is required to remove the electron [1 mark].*

2) a) *Group 3 [1 mark]*
There are three electrons removed before the first big jump in energy.

b) *The electrons are being removed from an increasing positive charge [1 mark] so more energy is needed to remove an electron / the force of attraction that has to be broken is greater [1 mark].*

c) *When an electron is removed from a different shell there is a big increase in the energy required (since that shell is closer to the nucleus) [1 mark].*

d) *There are 3 shells (because there are 2 big jumps in energy) [1 mark].*

Page 34 — Periodic Properties

1) *Mg has more delocalised electrons per atom [1 mark] and the ion has a greater charge density due to its smaller radius (because of the greater nuclear charge) [1 mark]. This gives Mg a stronger metal-metal bond, resulting in a higher boiling point [1 mark].*

2) a) *Increasing number of protons means a stronger pull from the positively charged nucleus [1 mark] making it harder to remove an electron from the outer shell [1 mark]. There are no extra inner electrons to add to the shielding effect [1 mark].*

b) (i) *Boron has the configuration $1s^2 2s^2 2p^1$ compared to $1s^2 2s^2$ for beryllium [1 mark]. The 2p shell is at a slightly higher energy level than the 2s shell. As a result, the extra distance and partial shielding of the 2s orbital make it easier to remove the outer electron [1 mark].*

(ii) *Electron repulsion in the shared 2p sub-shell [1 mark] makes it easier to remove an electron from oxygen than from nitrogen [1 mark]. (Oxygen has the configuration $1s^2 2s^2 2p^4$ compared to $1s^2 2s^2 2p^3$ for nitrogen.)*

3) *Neon has the configuration $1s^2 2s^2 2p^6$ and sodium $1s^2 2s^2 2p^6 3s^1$. [1 mark] The extra distance of sodium's outer electron from the nucleus and the extra electron shielding make it easier to remove than one of neon's 2p electrons [1 mark].*

4) a) *Si has a macromolecular (or giant molecular) structure [1 mark] consisting of very strong covalent bonds [1 mark].*

b) *Sulfur (S_8) has a larger molecule than phosphorus (P_4) [1 mark]. which results in larger London/van der Waals forces of attraction between molecules [1 mark].*

Unit 1: Section 4 — Bonding

Page 37 — Ionic Bonding

1) a)

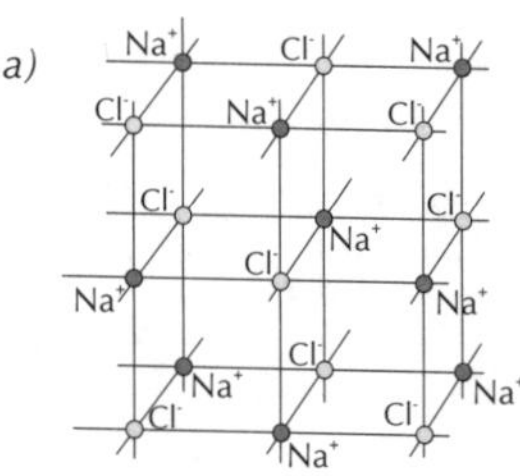

Your diagram should show the following —
- *cubic structure with ions at corners [1 mark]*
- *sodium ions and chloride ions labelled [1 mark]*
- *alternating sodium ions and chloride ions [1 mark]*

b) *giant ionic/crystal (lattice) [1 mark]*

c) *You'd expect it to have a high melting point [1 mark]. Because there are strong bonds between the ions [1 mark] due to the electrostatic forces [1 mark]. A lot of energy is required to overcome these bonds [1 mark].*

2) a) *Electrons move from one atom to another [1 mark].*
E.g. Na^+ [1 mark] Cl^- [1 mark]. Any correct examples of ions, one positive, one negative.

b) *In a solid, ions are held in place by strong ionic bonds [1 mark]. When the solid is heated to melting point, the ions gain enough energy [1 mark] to overcome the forces of attraction [1 mark] enough to become mobile [1 mark] and so carry charge (and hence electricity) through the substance [1 mark].*

3 a) *Sodium loses one (outer) electron to form Na^+ [1 mark]. Fluorine gains one electron to form F^- [1 mark]. Electrostatic forces of attraction between oppositely charged ions forms an ionic lattice [1 mark].*

b)

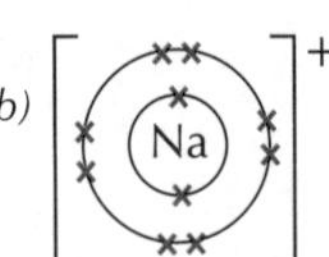

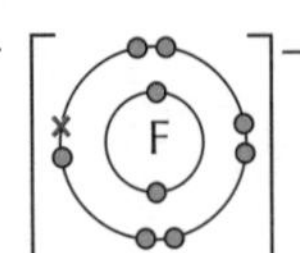

[1 mark for each correct structure. 1 mark for each correct charge.]

Answers

4) *High melting point suggests strong bonds must be overcome [1 mark]. Solubility in water suggests presence of ions [1 mark]. Fact that solution/liquid can conduct electricity suggests presence of ions [1 mark]. Solubility in water rules out giant covalent lattice [1 mark]. Non-conductivity of solid rules out metals [1 mark]. Any 4 marks max.*

Page 39 — More on Ionic Bonding

1) a) *Neon [1 mark]*

Neon has ten electrons. Oxygen has eight to begin with, and grabs two more to become O^{2-}. Sodium has eleven, and loses one to become Na^+.

b) *The O^{2-} ion has a larger ionic radius than the neon atom [1 mark]. The neon atom has two more protons than the O^{2-} ion, so it attracts its ten electrons more strongly [1 mark]. The Na^+ has a smaller ionic radius than the neon atom [1 mark]. The Na^+ ion has one more proton than the neon atom, so it attracts its ten electrons even more strongly [1 mark].*

2) a) *They are colourless [1 mark].*

Potassium ions have a 1+ charge. So the potassium ions in the diagram must be moving towards the cathode. Since you can't see a colourful streak moving in that direction, they must be colourless

b) *When a current is passed through the solution, the ions/particles move [1 mark] to the oppositely charged electrode [1 mark].*

Page 41 — Ions and Born-Haber Cycles

1) a) $\Delta H_f^{\ominus} = +177 + 590 + 1100 + 249 - 141 + 790 - 3401$

[2 marks. Deduct 1 mark for each error.]

$= -636$ kJ mol^{-1} *[1 mark]*

b) *Energy is required [1 mark] for the addition of the electron to the O^- ion due to the repulsion between the electron and the negatively charged O^- ion [1 mark].*

2) $\Delta H_{latt}^{\ominus} = -641 - 148 - 738 - 1451$ *[1 mark]* $- (122 \times 2)$ *[1 mark]* $- (-349 \times 2)$ *[1 mark]* $= -2524$ kJ mol^{-1} *[1 mark]*

Page 43— Lattice Energies & Polarisation of Ions

1) *Al^{3+} has a high charge/volume ratio (or a small radius AND a large positive charge) [1 mark], so it has a high polarising ability [1 mark] and can pull electron density away from Cl^- [1 mark] to create a bond with mostly covalent characteristics [1 mark]. (Alternatively Cl^- is relatively large [1 mark] and easily polarised [1 mark] so its electrons can be pulled away from Cl^- [1 mark] to create a bond with mostly covalent characteristics [1 mark].)*

2) a) *Increasing covalent character: NaBr, $MgBr_2$, MgI_2 [1 mark]. Covalent character is greatest when cations are small and have large charge, which applies more to Mg^{2+} than to Na^+ [1 mark], and when anions are large, which applies more to I^- than to Br^- [1 mark].*

b) *Experimental and theoretical lattice enthalpies match well when a compound has a high degree of ionic character [1 mark]. NaI has a higher degree of ionic character than MgI_2 because Na^+ has a smaller charge density / smaller charge and isn't much smaller than Mg^{2+} [1 mark].*

Page 45 — Covalent Bonding

1) a) *Covalent [1 mark]*

b)

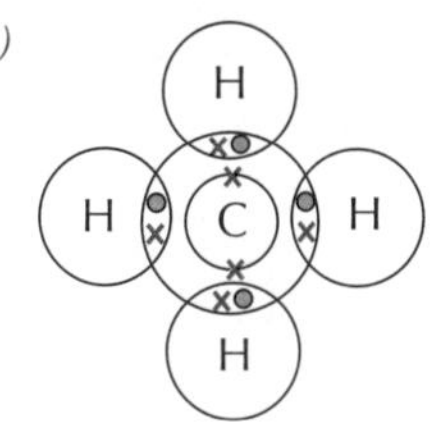

Your diagram should show the following —
- *a correct electron arrangement [1 mark]*
- *all 4 overlaps correct (one dot + one cross in each) [1 mark]*

2) *Ethene has a (C=C) double bond, made up of a σ and a π bond [1 mark]. Ethane only has a σ bond [1 mark]. The π bond is less tightly bound to the nuclei than the σ bond, so it is more reactive [1 mark].*

3) a)

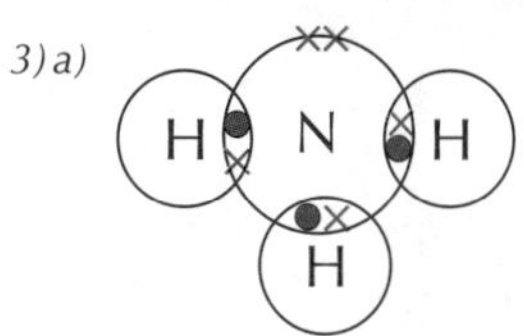

[1 mark for nitrogen and hydrogens covalently bonded. 1 mark for nitrogen's lone pair shown.]

b)

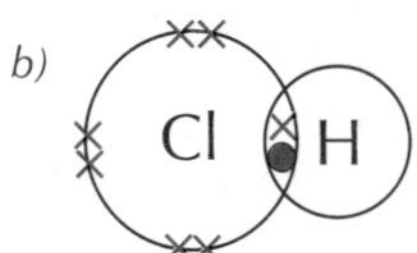

[1 mark for covalent bond. 1 mark for three lone pairs.]

c)

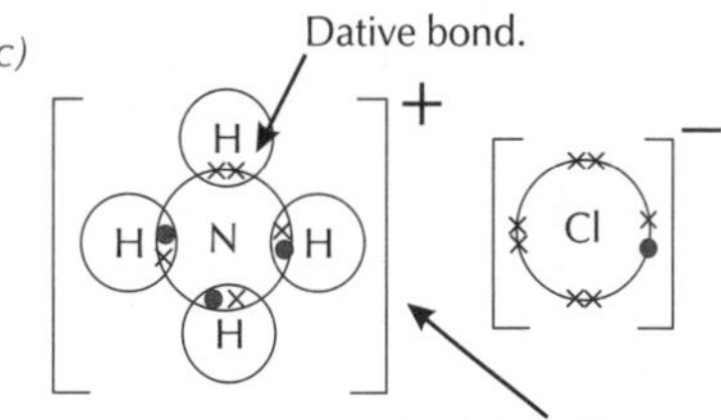

[1 mark for dative bond in ammonium ion, 1 mark for negative chloride ion, 1 mark for positive ammonium ion.]

Page 47 — Giant Covalent Structures & Metallic Bonding

1) a)

delocalised electron sea

lattice of +ve metal ions

[1 mark for showing any closely packed metal ions and 1 mark for showing a sea of delocalised electrons.]
Metallic bonding is the attraction between positive metal ions and a sea of delocalised electrons between them [1 mark].

2) *Bond type: covalent [1 mark], because its non-conductivity/ insolubility in water suggests no free electrons/free ions [1 mark]. Structure: giant network/lattice [1 mark], because its high melting point suggests that strong covalent bonds rather than weak intermolecular forces are being broken for the solid to melt [1 mark].*

3) a) *Increasing melting point: K, Na, Al [1 mark if Al highest, 2 marks for all three correct].*

b) *Aluminium has more delocalised electrons per atom (3) than sodium or potassium ions (both 1) [1 mark] so has the strongest metallic bonding. Sodium ions have a higher charge density / charge to volume ratio than potassium ions [1 mark] because both ions have a 1+ charge but the sodium ion is smaller [1 mark], so sodium has stronger bonding than potassium.*

Answers

Unit 1: Section 5 — Organic Chemistry

Page 49 — Organic Groups

1) a)

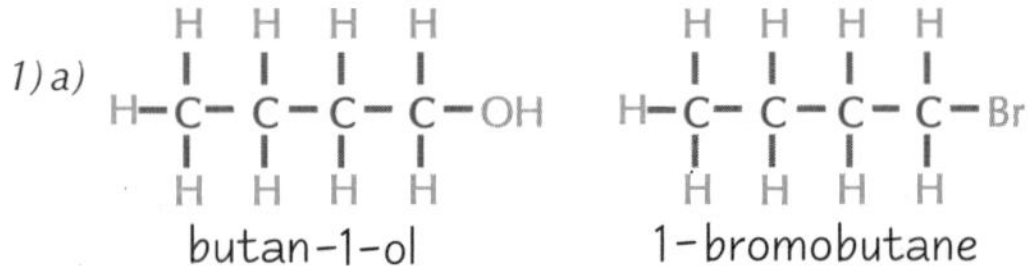

[1 mark for each correct structure]

b) –OH (hydroxyl) [1 mark].
It could be attached to the first or second carbon OR butan-2-ol also exists OR because the position of the –OH group affects its chemistry [1 mark].

c) Alcohols [1 mark]

2) a) i) 1-chloro-2-methylpropane
[2 marks available, lose 1 mark for each mistake]
Remember to put the substituents in alphabetical order.
ii) 3-methylbut-1-ene [2 marks available, lose 1 mark for each mistake]
iii) 2,4-dibromo-but-1-ene
[2 marks available, lose 1 mark for each mistake]
In parts ii) and iii), the double bond is the most important functional group, so it's given the lowest number.

b) (i) C_7H_{16}
(ii) $CH_3CH_2CH(CH_2CH_3)CH_2CH_3$

Page 51 — Alkanes and Structural Isomerism

1) a)

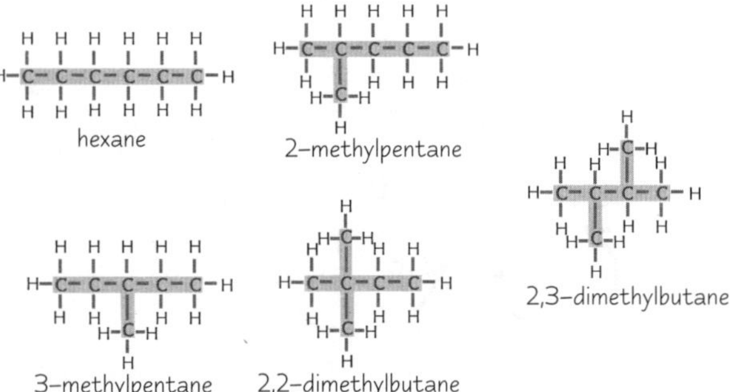

[1 mark for each correctly drawn isomer,
1 mark for each correct name]

b) alkanes [1 mark]

2) a) One with no double bond OR the maximum number of hydrogens OR single bonds only [1 mark]. It contains only hydrogen and carbon atoms [1 mark].

b) $C_2H_{6(g)} + 3\frac{1}{2}O_{2(g)} \rightarrow 2CO_{2(g)} + 3H_2O_{(g)}$
[1 mark for correct symbols, 1 mark for balancing]

c) $CH_3CH_3 + Br_2 \xrightarrow{U.V.} CH_3CH_2Br + HBr$ [1 mark]

Initiation: $Br_2 \xrightarrow{U.V.} 2Br\cdot$ [1 mark]

Propagation: $CH_3CH_3 + Br\cdot \rightarrow CH_3CH_2\cdot + HBr$ [1 mark]

$CH_3CH_2\cdot + Br_2 \rightarrow CH_3CH_2Br + Br\cdot$ [1 mark]

Termination: $CH_3CH_2\cdot + Br\cdot \rightarrow CH_3CH_2Br$

Or: $CH_3CH_2\cdot + CH_3CH_2\cdot \rightarrow CH_3CH_2CH_2CH_3$ [1 mark]

[1 mark for mentioning U.V.]
It's a free-radical [1 mark] substitution [1 mark] reaction.

Page 53 — Petroleum

1) a)(i) There's greater demand for smaller fractions [1 mark] for motor fuels [1 mark]. Or alternatively: There's greater demand for alkenes [1 mark] to make petrochemicals/polymers [1 mark].

(ii) E.g. $C_{12}H_{26} \rightarrow C_2H_4 + C_{10}H_{22}$ [1 mark].
There are loads of possible answers — just make sure the C's and H's balance and there's an alkane and an alkene.

b)(i) Cycloalkanes and arenes/aromatic hydrocarbons [1 mark for each].

(ii) They promote efficient combustion/reduce knocking (autoignition) [1 mark].

c)

2-methylbutane 2, 2-dimethylpropane

[1 mark for each structure, 1 mark for each name]

Page 55 — Fuels and Climate Change

1) a) $C_5H_{12} + 8O_2 \rightarrow 5CO_2 + 6H_2O$ [1 mark for reactants and products correct, 1 mark for correct balancing)

b) Both carbon dioxide and water (vapour) absorb infrared radiation [1 mark] that is emitted from the earth [1 mark]. They re-emit heat in all directions, including towards Earth, keeping the Earth warmer than it would otherwise be [1 mark].

c) The products of incomplete combustion include carbon monoxide gas / hydrocarbon/carbon particles) [1 mark] which is toxic / cause respiratory problems in humans [1 mark]

2) a) There is a limited amount of the resources/ they will run out eventually [1 mark].

b) Carbon dioxide is a greenhouse gas [1 mark]. Its increasing levels are resulting in an increase in average global temperatures [1 mark] and causing changes in climate / alterations in plant and animal habitats / increased pressure on food resources / more extreme weather conditions e.g. flooding, droughts. [2 marks for any two valid negative impacts].

Page 57 — Alkenes, Hazards and Risks

1) a) $C_{10}H_{16}$ [1 mark]

b) Three [1 mark]
The equivalent alkane would be $C_{10}H_{22}$, and each double bond removes 2 hydrogens.

c)

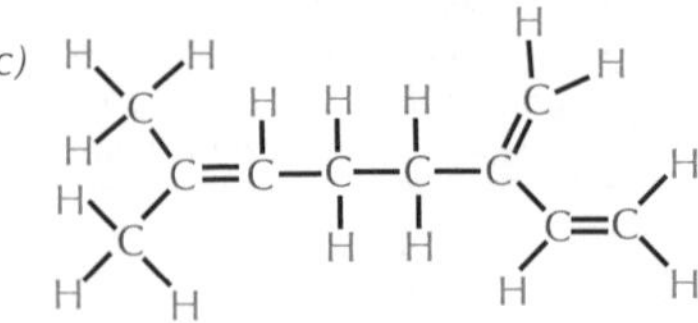

[2 marks. Lose 1 mark for an error.]

2) E.g. Keep away from naked flames / Use a fume cupboard / Work on a smaller scale / Use different, safer chemicals / Use lower concentrations [1 mark for each, up to a maximum of 2 marks].

Answers

Page 59 — Reactions of Alkenes

1) a) *Shake the alkene with bromine water [1 mark], and the solution goes colourless if a double bond is present [1 mark].*
   b) *Bromobutanol / 1-bromobutan-2-ol / 2-bromobutan-1-ol [1 mark]*
   c) *Electrophilic [1 mark] addition [1 mark].*
   d) (i)

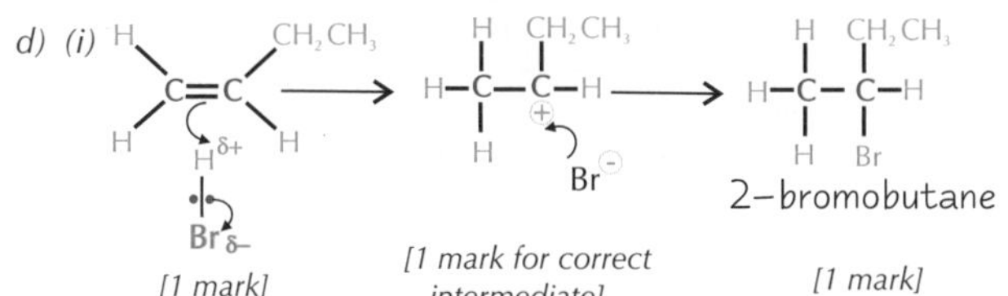

[1 mark] *[1 mark for correct intermediate]* *[1 mark]*

Check that your curly arrows are exactly right, or you'll lose marks. They have to go from exactly where the electrons are from, to where they're going to.

   (ii) *The secondary carbocation OR the carbocation with the most attached alkyl groups [1 mark] is the most stable intermediate and so is the most likely to form [1 mark].*

Page 61 — E/Z Isomerism

1) a) E-pent-2-ene *[1 mark]* *[1 mark]* Z-pent-2-ene *[1 mark]* *[1 mark]*
   b) *E/Z isomers occur because atoms can't rotate about C=C double bonds [1 mark]. Alkenes contain C=C double bonds and alkanes don't, so alkenes can form E/Z isomers and alkanes can't [1 mark].*
2) 2 *[1 mark]*

Page 63 — Polymers

1) a) *[1 mark]*
   b) *[1 mark]*
2) a) *Saves on landfill OR Energy can be used to generate electricity [1 mark for either]*
   b) *Toxic gases produced [1 mark]. Scrubbers can be used [1 mark] to remove these toxic gases.*
3) *Renewable raw material / Less energy used (in manufacture) / Less CO_2 produced (over lifetime of polymer) [1 mark for each, up to a maximum of 2 marks]*

Unit 2: Section 1 — Bonding & Intermolecular Forces

Page 65 — Shapes of Molecules

1) a) NCl_3 *[1 mark]* BCl_3 *[1 mark]*
   b) NCl_3 *[1 mark]*
   shape: trigonal pyramidal [1 mark], bond angle: 107° (accept between 105° and 109°) [1 mark]
   BCl_3 *[1 mark]*
   (It must be a reasonable "Y" shaped molecule.)
   shape: trigonal planar [1 mark], bond angle: 120° exactly [1 mark]
   c) *BCl_3 has three electron pairs around the central B atom. [1 mark] NCl_3 has four electron pairs around N [1 mark], including one lone pair. [1 mark]*

Page 67 — Carbon Structures

1) a) (i) *In graphite each carbon atom is bonded to three other carbon atoms [1 mark], which leaves one electron free to move through the solid and so carry a charge [1 mark]. In diamond, all four outer shell electrons are involved in covalent bonds [1 mark] so there are no free electrons to carry charge [1 mark].*
   (ii) *In diamond all four bonds are C-C covalent bonds i.e. the distance between all atoms is one C-C bond length [1 mark]. Graphite forms layers and the distance between layers is greater than a C-C bond length [1 mark].*
   (iii) *The forces between layers in graphite are weak intermolecular bonds [1 mark] so the layers can slide over each other [1 mark], whereas in diamond all the bonds are strong covalent bonds [1 mark].*
   b) *Both have high melting (subliming) points / Both are insoluble in all/most solvents [1 mark for any correct shared physical property].*
2) a) *Sixty carbon atoms [1 mark], with each carbon atom bonded to three other carbon atoms [1 mark], to form a hollow ball [1 mark].*
   b) *E.g. any one from: Small wires in circuits [1 mark] — the structure contains free electrons that can conduct electricity [1 mark]. / Making building materials or sports equipment [1 mark] — network of covalent bonds gives a very strong material, that is also light since the molecules are hollow [1 mark]. / Drug delivery mechanism [1 mark] — the hollow tube can be used to cage another molecule, such as a drug molecule, and carry it to the cells of the body [1 mark].*

Page 69 — Electronegativity and Polarisation

1) a) *The power of an atom to withdraw electron density [1 mark] from a covalent bond [1 mark] OR the ability of an atom to attract the bonding electrons [1 mark] in a covalent bond [1 mark].*
   b) (i) Br — Br non-polar (or no dipole)
   (ii) polar (or dipole)
   (iii) non-polar (or no dipole)
   (iv) polar (or dipole)

   [For each molecule: 1 mark for correct shape and bond polarities, 1 mark for correct overall polarity].

   To help you decide if the molecule's polar or not, imagine the atoms are having a tug of war with the electrons. If they're all pulling the same amount in opposite directions, the electrons aren't going to go anywhere.
   c) *The lone pair of electrons on nitrogen [1 mark] cancels out the dipole or polarity [1 mark].*
   This can't happen with NH_3 because the dipole's in the opposite direction.

Answers

Page 71 — Intermolecular Forces

1) a) *London OR van der Waals OR instantaneous/temporary dipole-induced dipole OR dispersion forces.*
Permanent dipole-dipole interactions/forces.
Hydrogen bonding.
Permanent dipole-induced dipole interactions.
[1 mark each for any three]

b)

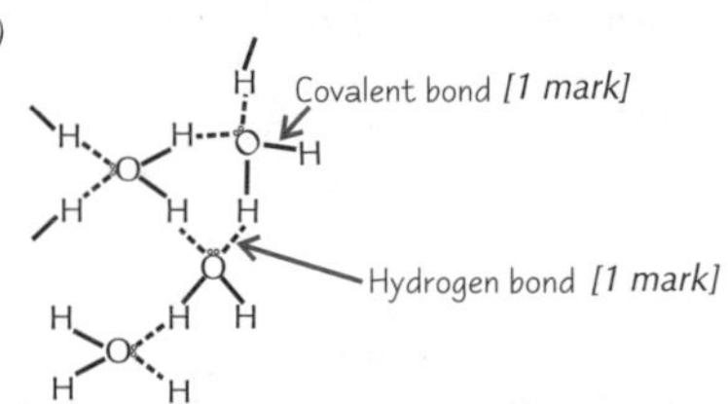

You could have shown the H_2O molecules in either of these two ways. *[1 mark]*

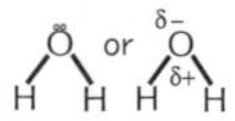

And London OR van der Waals OR instantaneous/temporary dipole-induced dipole OR dispersion forces of attraction between water molecules. [1 mark]

c) *The increase in the size/M_r and the number of electrons [1 mark] leads to an increase in London / van der Waals / instantaneous/ temporary dipole-induced dipole forces [1 mark]. Water has a higher boiling point than would be expected from the general trend because there is hydrogen bonding between water molecules but not in any of the other substances [1 mark] and hydrogen bonding is a stronger force than London forces [1 mark].*

2) a) (i) 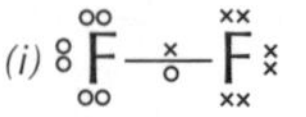 F—F *[1 mark]*

(ii) 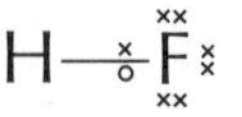 H—F

[1 mark. Bonding electrons may be shown in the centre of bond.]

b) *There must be a difference in electronegativity values if there is to be a permanent dipole [1 mark] and this is only the case for HF [1 mark].*

c) *HF is liquid at room temperature. Because HF is polar / has a permanent dipole [1 mark], there are stronger intermolecular forces between molecules so it has a higher boiling point [1 mark].*

Page 73 — Solubility

1) a)(i) *Hydrogen bonds [1 mark] form between the alcohol and water molecules [1 mark].*
The (hydrogen) bonds between water molecules are stronger [1 mark] than bonds that would form between water and the halogenoalkane molecules [1 mark].
OR
The halogenoalkanes do not contain strong enough dipoles [1 mark] to form hydrogen bonds with water [1 mark].

(ii)

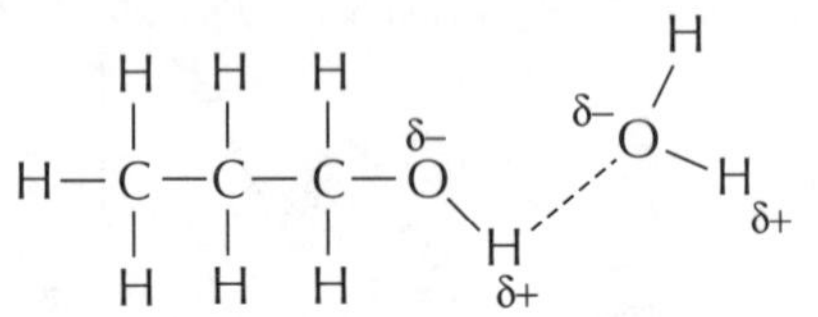

[1 mark for the two substances with relevant δ+ and δ– marked correctly, 1 mark for showing the bond between the hydrogen of the propanol molecule and oxygen of the water molecule.]

b) *K^+ ions are attracted to the δ– ends of the water molecules [1 mark] and I^- ions are attracted to the δ+ ends [1 mark]. This overcomes the ionic bonds in the lattice / The ions are pulled away from the lattice [1 mark], and surrounded by water molecules [1 mark], forming hydrated ions:*

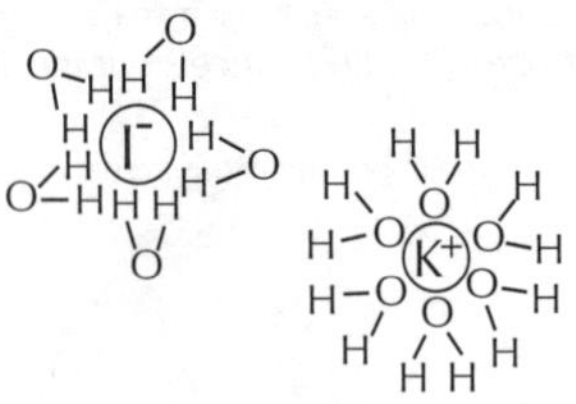

[1 mark].

2) a) *Try to dissolve the substance in water [1 mark] and hexane (or other non-polar solvent) [1 mark]*
If X is non-polar, it is likely to dissolve in hexane [1 mark], but not in water [1 mark].
Remember 'like dissolves like' — in other words, substances usually dissolve best in solvents that have similar bonding.

b) *X and hexane have London/van der Waals forces between their molecules [1 mark] and form similar bonds with each other [1 mark]. Water has hydrogen bonds [1 mark] which are much stronger than the bonds it could form with a non-polar compound [1 mark].*

Unit 2: Section 2 — Inorganic Chemistry

Page 75 — Oxidation and Reduction

1) a) $H_2SO_{4\,(l)} + 8HI_{(g)} \rightarrow H_2S_{(g)} + 4I_{2\,(s)} + 4H_2O_{(l)}$ *[1 mark]*

b) *Ox. No. of S in H_2SO_4 = +6 [1 mark]*
Ox. No. of S in H_2S = -2 [1 mark]

c) $2I^- \rightarrow I_2 + 2e^-$ *[1 mark]*

d) $H_2SO_4 + 8H^+ + 8e^- \rightarrow H_2S + 4H_2O$
[all species correct — 1 mark, balancing — 1 mark]

e) *Iodide [1 mark] — it donates electrons / its oxidation number increases [1 mark]*
The ionic equations here are pretty tricky. Use the equation you're given as much as possible. For part d), sulfur is being reduced from +6 to –2, so it's gaining 8 electrons. You also need to add H^+s and H_2O's to balance it. With ionic equations, always make sure the charges balance. E.g. in part d), charge on left = +8 – 8 = 0 = right-hand side.

2) a) (i) $FeSO_4$ (ii) $Fe(OH)_2$ (iii) $Fe_2(SO_4)_3$ (iv) $Fe(OH)_3$
[1 mark for each]

b) *The oxygen in the atmosphere would oxidise iron(II) compounds to iron(III) compounds. [1 mark]*

Page 77 — Group 2

1) a) *Wind/burping etc. [1 mark]*

b) *Magnesium hydroxide / magnesium oxide [1 mark]*

c) $Mg(OH)_2 + 2HCl \rightarrow MgCl_2 + 2H_2O$ *[1 mark]*
$MgO + 2HCl \rightarrow MgCl_2 + H_2O$ *[1 mark]*

2) a) $Ca_{(s)} + Cl_{2(g)} \rightarrow CaCl_{2(s)}$ *[1 mark]*

b) *From 0 to +2 [1 mark]*

c) *White [1 mark] solid [1 mark]*

d) *Ionic [1 mark]*
...because as everybody who's anybody knows, Group 2 compounds (including oxides) are generally white ionic solids.

Page 79 — Group 1 and 2 Compounds

1) **A** = CO_2 *[1 mark] (it turns limewater cloudy)*
B = CaO *[1 mark] (CO_2 is released when a carbonate is heated, leaving an oxide)*
Original compound = $CaCO_3$ *[1 mark] (CO_2 is released when a carbonate is heated, so the original compound must have been calcium carbonate)*
...the easiest one to get is gas A because it's just describing the limewater test for CO_2. Getting the others isn't hard either, but you've got to really know all the equations from page 78 — or you'll get muddled.

2) a) $2NaNO_{3(s)} \rightarrow 2NaNO_{2(s)} + O_{2(g)}$ *[1 mark]*

b) *O_2 gas relights a glowing splint. [1 mark]*

c) *magnesium nitrate sodium nitrate potassium nitrate [1 mark]*
Group 2 nitrates decompose more easily than Group 1 (the greater the charge on the cation, the less stable the nitrate anion) [1 mark].
The further down the group, the more stable the nitrate (the larger the cation, the less distortion to the nitrate anion) [1 mark].

3) a) *Energy is absorbed and electrons move to higher energy levels. [1 mark] Energy is released in the form of coloured light when the electrons fall back to the lower levels [1 mark].*

b) *caesium [1 mark]*

Answers

Page 81 — The Halogens

1) a) $2OH^- + Cl_2 \rightarrow OCl^- + Cl^- + H_2O$ [1 mark]
   b) Disproportionation is simultaneous oxidation and reduction of an element in a reaction [1 mark]. Cl_2 has been reduced to Cl^- [1 mark] and oxidised to OCl^- [1 mark].
2) a) MgF_2 [1 mark] (oxidation state of fluoride = -1)
   b) KBrO [1 mark] (oxidation state of bromine in bromate(I) = +1, and the name brom**ate** indicates oxygen is present)
   c) $NaClO_3$ [1 mark] (oxidation state of chlorine in chlorate(V) = +5, and the name chlor**ate** indicates oxygen is present)
3) a) copper(II) chloride [1 mark]
   b) Oxidation step: $Cu \rightarrow Cu^{2+} + 2e^-$
   Reduction step: $Cl_2 + 2e^- \rightarrow 2Cl^-$
   [3 marks — 1 mark for each correct equation and 1 mark for correctly labelling them as oxidation and reduction]

Page 84 — Reactions of the Halides

1) **Aqueous** solutions of both halides are tested [1 mark].
   a) **Sodium chloride** — silver nitrate gives white precipitate which dissolves in dilute ammonia solution [1 mark].
   $Ag^+ + Cl^- \rightarrow AgCl$ [1 mark]
   Sodium bromide — silver nitrate gives cream precipitate which is only soluble in concentrated ammonia solution [1 mark].
   $Ag^+ + Br^- \rightarrow AgBr$ [1 mark]
   b) **Sodium chloride** — Misty fumes [1 mark]
   $NaCl + H_2SO_4 \rightarrow NaHSO_4 + HCl$ [1 mark]
   Sodium bromide — Misty fumes [1 mark]
   $NaBr + H_2SO_4 \rightarrow NaHSO_4 + HBr$ [1 mark]
   $2HBr + H_2SO_4 \rightarrow Br_2 + SO_2 + 2H_2O$ [1 mark]
   Orange / brown vapour [1 mark]
2) a) KI (via HI) reduces H_2SO_4 to H_2S [1 mark]. The reducing power of halide ions increases down the group [1 mark] and At is below I in the group [1 mark], so H_2S will be produced [1 mark].
   b) AgI is insoluble in concentrated ammonia solution [1 mark].
   The solubility of halides in ammonia solution decreases down the group [1 mark], so AgAt will **NOT** dissolve. [1 mark]
   Question 2 is the kind of question that could completely throw you if you're not really clued up on the facts. If you really know p82, then in part a) you'll go, "Ah - ha!!! Reactions of halides with H_2SO_4 — reducing power increases down the group..." If not, you basically won't have a clue. The moral is... it really is just about learning all the facts. Boring, but true.
3) a) This is a redox reaction because there are changes in oxidation numbers / oxidation and reduction happen simultaneously (Cu goes from +2 to +1 and I from –1 to 0) / electrons are transferred [1 mark].
   The reducing agent is iodide ions (which are themselves oxidised) [1 mark].
   b) The potassium iodide solution would go from colourless [1 mark] to a deep red/brown solution [1 mark].
   c) $Cl_{2\,(aq)} + 2NaI_{(aq)} \rightarrow 2NaCl_{(aq)} + I_{2\,(aq)}$
   [1 mark for correct products and reactants, 1 mark for correct balancing (can also leave out Na spectator ion)].
   Chlorine is the oxidising agent [1 mark].

Page 87 — Acid-Base Titrations and Uncertainty

1) a) You can make the data more reliable by repeating the titration several more times and using the mean [1 mark].
   b) The titre is calculated by subtracting the initial volume from the final volume. Each of these has an uncertainty of 0.05 cm^3, so the total uncertainty is 0.1 cm^3.
   percentage uncertainty = (0.1/3.1) × 100 = **3.23%**
   [1 mark for correct use of percentage uncertainty formula, 1 mark for using uncertainty of 0.1 cm^3]
   c) The percentage error will decrease if the titres are larger [1 mark].
   Using a less concentrated solution will result in larger titres [1 mark].
2) a) no. moles of HCl = 0.1 mol dm^{-3} × (19.25 cm^3/1000) = 0.001925 [1 mark]
   no. moles of NaOH = 0.001925 (since reacting ratio is 1:1) [1 mark]
   concentration of NaOH = 0.001925/(25 cm^3/1000)
   = **0.077 mol dm^{-3}** [1 mark]
   b) percentage uncertainty in pipette
   = (0.06/25.0) × 100 = 0.24% [1 mark]
   percentage uncertainty in titre = (0.1/19.25) × 100 = 0.52% [1 mark]
   Total percentage uncertainty = 0.76% [1 mark]
   So uncertainty of concentration = 0.76% of 0.077
   = **0.000585 mol dm^{-3} (5.85 × 10^{-4} mol dm^{-3})** [1 mark]

Page 89 — Iodine-Sodium Thiosulfate Titration

1) a) $IO_3^- + 5I^- + 6H^+ \rightarrow 3I_2 + 3H_2O$
   [1 mark for correct reactants and products and 1 mark for balancing]
   b) Number of moles of thiosulfate = 0.15 × (24 ÷ 1000)
   = 3.6 × 10^{-3} moles [1 mark]
   c) 2 moles of thiosulfate reacts with 1 mole of iodine, so there were (3.6 × 10^{-3}) ÷ 2 = 1.8 × 10^{-3} moles of iodine [1 mark]
   d) 1/3 mole of iodate(V) ions produces 1 mole of iodine molecules [1 mark]
   e) There must be 1.8 × 10^{-3} ÷ 3 = 6 × 10^{-4} moles of iodate(V) solution [1 mark].
   So concentration of potassium iodate(V) =
   6 × 10^{-4} moles × 1000 ÷ 10 = 0.06 mol dm^{-3} [1 mark].

Unit 2: Section 3 — Kinetics and Equilibria

Page 91 — Reaction Rates

1) The molecules don't always have enough energy [1 mark].
   Collisions don't always happen in the right orientation (the molecules mightn't be facing each other in the best way and will just bounce off each other) [1 mark].
2) The particles in a liquid move freely and all of them are able to collide with the solid particles [1 mark]. Particles in solids just vibrate about fixed positions, so only those on the touching surfaces between the two solids will be able to react [1 mark].
3) a) X
   The X curve shows the same total number of molecules as the 25°C curve, but more of them have lower energy.
   b) The shape of the curve shows fewer molecules [1 mark] have the required activation energy [1 mark].

Page 93 — Catalysts and Reaction Rate Experiments

1) a)

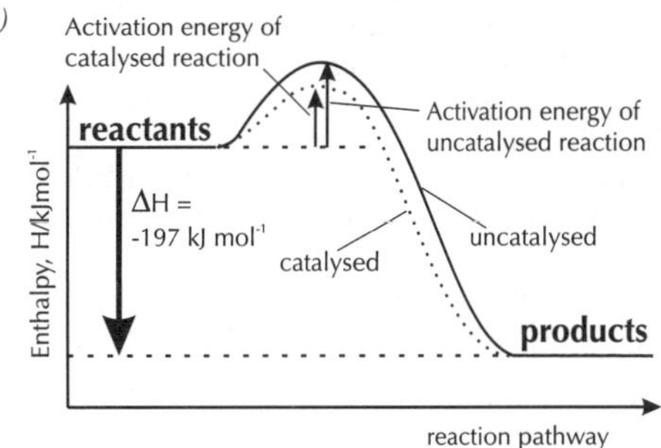

   Curve showing activation energy [1 mark]. This must link reactants and products. Showing exothermic change (products lower in energy than reactants), with ΔH correctly labelled and a downward arrow [1 mark]. Correctly labelling activation energy (from reactants to highest energy peak) [1 mark].
   Label your axes correctly. (No, not the sharp tools for chopping wood or heads off — you know what I mean.)
   b) See the diagram above. Reaction profile showing a greater activation energy than for the catalysed reaction [1 mark].
   Remember — catalysts lower the activation energy. So uncatalysed reactions have greater activation energies.
   c) A catalyst increases the rate of the reaction by providing an alternative reaction pathway [1 mark], with a lower activation energy [1 mark].

Answers

2) a) $2H_2O_{2(l)} \rightarrow 2H_2O_{(l)} + O_{2(g)}$

Correct symbols [1 mark] and balancing equation [1 mark]. You get the marks even if you forgot the state symbols.

b)

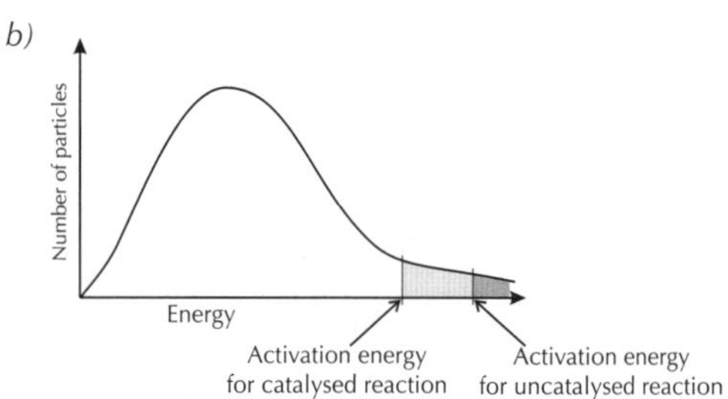

Correct general shape of the curve [1 mark]. Correctly labelling the axes [1 mark]. Activation energies marked on the horizontal axis — the catalysed activation energy must be lower than the uncatalysed activation energy [1 mark].

You don't have to draw another curve for the catalysed reaction. Just mark the lower activation energy on the one you've already done.

c) *Manganese(IV) oxide lowers the activation energy by providing an alternative reaction pathway [1 mark]. So more reactant molecules have energies greater than the activation energy [1 mark], meaning there are more successful collisions in a given period of time, and so the rate increases [1 mark].*

Page 95 — Chemical Equilibria

1) a) (i) *There's no change [1 mark]. There's the same number of molecules/moles on each side of the equation [1 mark].*

(ii) *Reducing temperature removes heat. So the equilibrium shifts in the exothermic direction to release heat [1 mark]. The reverse reaction is exothermic (since the forward reaction is endothermic). So, the position of equilibrium shifts left [1 mark].*

(iii) *Removing nitrogen monoxide reduces its concentration. The equilibrium position shifts right to try and increase the nitrogen monoxide concentration again [1 mark].*

b) *No effect [1 mark].*

Catalysts don't affect the equilibrium position. They just help the reaction to get there sooner.

Page 97 — More About Equilibria

1) a) *At low temperature the particles move more slowly [1 mark]. This means fewer successful collisions [1 mark] and a slower rate of reaction [1 mark].*

b) *High pressure is expensive. [1 mark] The cost of the extra pressure is greater than the value of the extra yield. [1 mark]*

c) (i) *The rate of production of ethanol is increased [1 mark] because the catalyst provides an alternative reaction route [1 mark] of lower activation energy [1 mark]*

(ii) *The amount of ethanol in the equilibrium mixture stays the same [1 mark] because a catalyst has no effect on the position of equilibrium [1 mark]*

You don't need to know anything about how ethanol is made. The general principles are the same for any reaction.

Unit 2: Section 4 — Organic Chemistry 2

Page 99 — Alcohols

1) a) *primary:*

```
   H  H  H  H  H
   |  |  |  |  |
H—C—C—C—C—C—OH
   |  |  |  |  |
   H  H  H  H  H
```

[1 mark], pentan-1-ol [1 mark]

secondary:

```
   H  H  H  H  H
   |  |  |  |  |
H—C—C—C—C—C—H
   |  |  |  |  |
   H  H  H  OH H
```

[1 mark], pentan-2-ol [mark].

tertiary:

```
            H
            |
          H—C—H
   H  H     |   H
   |  |     |   |
H—C—C—C—C—H
   |  |     |   |
   H  H    OH   H
```

[1 mark],

2-methylbutan-2-ol [1 mark]

b) (i) *React ethanol with phosphorus(III) bromide, PBr_3 [1 mark]. Prepare PBr_3 in situ by reacting Br with (red) phosphorus [1 mark].*

(ii) *React ethanol with excess sodium metal [1 mark].*

2) a) $CH_3CH(OH)CH_3 + PCl_5 \rightarrow CH_3CH(Cl)CH_3 + HCl + POCl_3$

[1 mark for LHS correct, 1 mark for RHS correct]

b) *Carry out the reaction in a fume cupboard [1 mark]*

3) a) $2CH_3CH_2CH_2CH_2OH + 2Na \rightarrow 2CH_3CH_2CH_2CH_2O^-Na^+ + H_2$

[1 mark for correct reactants and products, 1 mark for correct balancing]

b) *sodium 1-butoxide*

c) *More vigorous bubbling / the reaction with methanol would be more vigorous than with butan-1-ol [1 mark] because methanol's hydrocarbon chain is shorter [1 mark] and the reaction is faster with shorter-chain alcohols.*

Page 101 — Oxidation of Alcohols

1) a) *[1 mark for diagram]*

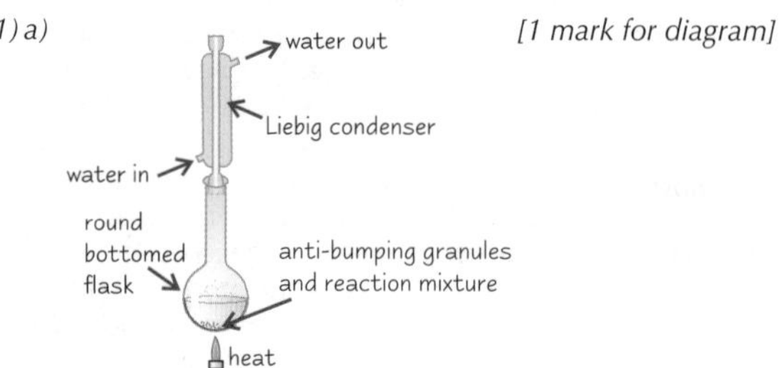

You set up reflux apparatus in this way so that the reaction can be heated to boiling point [1 mark] without losing any materials/reactants/products OR so vapour will condense and drip back into the flask [1 mark]

b) (i) *Warm with Fehling's/Benedict's solution: turns from blue to brick-red OR warm with Tollen's reagent: a silver mirror is produced [1 mark for test, 1 mark for result]*

(ii) *Propanoic acid [1 mark]*

(iii) $CH_3CH_2CH_2OH + [O] \rightarrow CH_3CH_2CHO + H_2O$ *[1 mark]*

$CH_3CH_2CHO + [O] \rightarrow CH_3CH_2COOH$ *[1 mark]*

(iv) *Distillation [1 mark]. This is so aldehyde is removed immediately as it forms [1 mark].*

If you don't get the aldehyde out quick-smart, it'll be a carboxylic acid before you know it.

c) (i)

```
        H
        |
      H—C—H
   H    |
   |    |
H—C—C—OH
   |    |
   H    |
      H—C—H
        |
        H
```

[1 mark]

(ii) *2-methylpropan-2-ol is a tertiary alcohol (which is more stable) [1 mark].*

Answers

Page 103 — Halogenoalkanes

1) a) 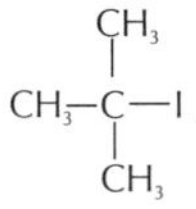 $CH_3-C(CH_3)_2-I$ *[1 mark] 2-methyl-2-iodopropane [1 mark]*

b) *AgI [1 mark]*

c) $CH_3CH_2CH_2CH_2I + H_2O \rightarrow CH_3CH_2CH_2CH_2OH + HI$ *[1 mark]*
Accept molecular formulas and $H^+ + I^-$ instead of HI.

2) a) 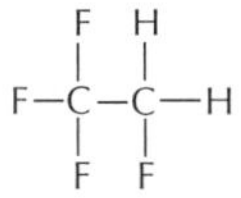 $F-CF_2-CHF-H$ (F–C(F)(F)–C(H)(F)–H) *[1 mark]*

b) *They are gases at room temperature. / They are unreactive. / They are easily compressible / They are non-toxic. [1 mark each for any two sensible answers]*

Page 105 — Reactions of Halogenoalkanes

1) a) ***Reaction 1***
Reagent — NaOH/KOH/OH^- [1 mark]
Solvent — Aqueous solution/water [1 mark]
Reaction 2
Reagent — Ammonia/NH_3 [1 mark]
Solvent — Ethanol/alcohol [1 mark]
Reaction 3
Reagent — NaOH/KOH [1 mark].
Solvent — Ethanol/alcohol [1 mark]

b) *(i) A cream-coloured precipitate would form [1 mark].*
(ii) $CH_3CHBrCH_{2\,(aq)} + H_2O_{(l)} + Ag^+_{(aq)} \rightarrow CH_3CH(OH)CH_{2\,(aq)} + H^+_{(aq)} + AgBr_{(s)}$.
[1 mark for reactants and products, 1 mark for state symbols]

Unit 2: Section 5 — Mechanisms, Spectra and Green Chemistry

Page 107 — Types of Reaction

1) a) *an atom or molecule with an unpaired electron [1 mark]*

b) *(i)* $Cl-Cl \rightarrow 2Cl\bullet$ *[1 mark]*
(ii) $Cl\bullet + CH_4 \rightarrow HCl + CH_3\bullet$ *[1 mark]*
(iii) Any two radicals combining, e.g. $Cl\bullet + CH_3\bullet \rightarrow CH_3Cl$ *[1 mark]*

2) a) *nucleophilic substitution [1 mark]*

b) *Heat/reflux [1 mark], (excess) ammonia [1 mark] in ethanol [1 mark]*

3) a) *ethanol [1 mark],* CH_3CH_2OH *[1 mark]*

b) *propan-2-ol [1 mark],* $CH_3CH(OH)CH_3$ *[1 mark]*

Page 109 — The Ozone Layer

1) a) *Ozone is formed by the effect of UV radiation from the sun on oxygen molecules. [1 mark] The oxygen molecules split to form oxygen free radicals [1 mark] which react with more oxygen molecules to form ozone [1 mark].*

b) *UV radiation can cause skin cancer. [1 mark] The ozone layer prevents most harmful UV radiation from the sun from reaching the Earth's surface. [1 mark]*

c) *The ozone molecules interact with UV radiation to form an oxygen molecule and a free oxygen radical* ($O_3 + h\nu \rightarrow O_2 + O\bullet$) *[1 mark]*
The radical produced then forms more ozone with an O_2 molecule. ($O_2 + O\bullet \rightarrow O_3$) *[1 mark]*

2) a) *Coolants in fridges / aerosol propellants / fire extinguishers / foaming plastics [1 mark each use, up to a maximum of 3 marks]*

b) *They are unreactive/chemically stable. They are non-flammable. They are non-toxic. They are volatile. [1 mark for each, up to a maximum of 3 marks]*

c) *Because they were destroying the ozone layer. [1 mark]*

3) a) *Nitrogen monoxide [mark].*
$NO\bullet + O_3 \rightarrow O_2 + NO_2\bullet$
$NO_2\bullet + O_3 \rightarrow 2O_2 + NO\bullet$
[1 mark for showing NO• reacting with an ozone molecule, 1 mark for showing NO• regenerated. 1 mark for balanced equations.]

b) *Motor vehicle/aircraft engines [1 mark]*

Page 111 — Green Chemistry

1) a) *If a catalyst is used, less energy is needed for the process [1 mark] more efficient processes are made possible [1 mark]*

b) *The process has 100% atom economy as only one product is made [1 mark] so no waste is produced [1 mark].*

Page 114 — Climate Change

1) a) *Water vapour, carbon dioxide, methane [2 marks for three correct answers, 1 mark for two]*

b) *The Earth emits infrared radiation/heat [1 mark], some of which is absorbed by greenhouse gases [1 mark]. The greenhouse gases re-emit infrared radiation in all directions, including towards Earth [1 mark].*

c) *How much radiation one molecule of the gas absorbs [1 mark]*
How much of the gas there is in the atmosphere [1 mark]

2 a) *Carbon dioxide is released, which was removed from the atmosphere millions of years ago [1 mark].*

b) *(i) Fermentation [1 mark]*
(ii) Every bit of carbon that is released into the atmosphere when the fuel is burned is removed [1 mark] when the next crop of sugar cane is grown / was removed by the crop as it grew [1 mark]

Page 117 — Mass Spectra and Infrared Spectra

1) a) *44 [1 mark]*

b) *X has a mass of 15. It is probably an methyl group/CH_3. [1 mark]*
Y has a mass of 29. It is probably a ethyl group/C_2H_5. [1 mark]

c) 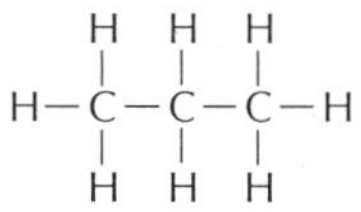 $H-CH_2-CH_2-CH_2-H$ (H–C(H)(H)–C(H)(H)–C(H)(H)–H) *[1 mark]*

d) *If the compound was an alcohol, you would expect a peak with m/z ratio of 17 [1 mark], caused by the OH fragment [1 mark].*

2 a) *A's due to an O–H group in a carboxylic acid [1 mark].*
B's due to a C=O as in an aldehyde, ketone or acid [1 mark].

b) *The spectrum suggests it's a carboxylic acid — it's got a COOH group [1 mark]. This group has a mass of 45, so the rest of the molecule has a mass of 29 (74 – 45), which is likely to be C_2H_5 [1 mark]. So the molecule could be C_2H_5COOH — propanoic acid [1 mark].*

Index

Index

Index